Edexcel
Physics
for AS

Graham George
Mike Benn

DYNAMIC LEARNING
Innovate • Motivate • Personalise
CD-ROM INSIDE

HODDER
EDUCATION
AN HACHETTE UK COMPANY

The Publishers would like to thank the following for permission to reproduce copyright material:
Photo credits **p.1** Science Photo Library/Gustoimages; **p.79** Science Photo Library/Mehau Kulyk; **p.98** Science Photo Library/Peter Aprahamian; **p.100** Science Photo Library/Sovereign, ISM; **p.114** *t, b* Science Photo Library/Andrew Lambert; **p.148** *t* Alamy/David J. Green, *b* Alamy/Science Photos; **p.156** *l* Science Photo Library/Victor Habbick Visions, *r* Science Photo Library/Martin Bond; **p.167** *t, b* Alamy/Leslie Garland Picture Library; **p.172** *t, b* Science Photo Library; **p.176** Science Photo Library/US Library of Congress; **p.180** Science Photo Library/US National Archives; **p.181** *t* Science Photo Library, *b* Science Photo Library/Emilio Segre Visual Archive; **p.182** Science Photo Library.
b = bottom, *l* = left, *r* = right, *t* = top

Every effort has been made to trace all copyright holders, but if any have been inadvertently overlooked the Publishers will be pleased to make the necessary arrangements at the first opportunity.

Although every effort has been made to ensure that website addresses are correct at time of going to press, Hodder Education cannot be held responsible for the content of any website mentioned in this book. It is sometimes possible to find a relocated web page by typing in the address of the home page for a website in the URL window of your browser.

Hachette UK's policy is to use papers that are natural, renewable and recyclable products and made from wood grown in sustainable forests. The logging and manufacturing processes are expected to conform to the environmental regulations of the country of origin.

Orders: please contact Bookpoint Ltd, 130 Milton Park, Abingdon, Oxon OX14 4SB. Telephone: (44) 01235 827720. Fax: (44) 01235 400454. Lines are open 9.00 – 5.00, Monday to Saturday, with a 24-hour message answering service. Visit our website at www.hoddereducation.co.uk

© Mike Benn, Graham George 2008
First published in 2008 by
Hodder Education,
An Hachette UK Company
338 Euston Road
London NW1 3BH

Impression number   5 4 3
Year   2012 2011 2010 2009

Cover photo © Leo Mason/Corbis
Illustrations by Ken Vail Graphic Design and Beehive Illustration (Laszlo Veres)
Index by Indexing Specialists (UK) Ltd
Typeset in Goudy 10.5pt by Ken Vail Graphic Design, Cambridge
Printed in Italy

A catalogue record for this title is available from the British Library

ISBN: 978 0340 88802 5

Contents

Introduction

Welcome to *Edexcel Physics for* AS. The Edexcel specification was developed from the best of the Edexcel concept-led and the Salters Horners context-led specifications. Between them, the authors have a wealth of experience of teaching and examining both these previous specifications. Although this book has been specifically written to cover the concept approach to the new specification, it would also be a very valuable source book for the context approach as it is illustrated throughout by contextual examples.

A key aspect of the book is the emphasis on practical work. Virtually all the experiments suggested in the specification are described in such a way that students can carry out the experiments in a laboratory environment. In addition, each experiment is illustrated with typical data for the reader to work through without requiring access to a laboratory. In the interests of Health & Safety, teachers are strongly advised to carry out an independent risk assessment for each experiment, if in doubt referring to an organisation such as *CLEAPSS* for advice.

As well as the practical activities, the text contains many worked examples for you to test yourself as you go along. At the end of each topic you will find exam-style Review questions to further help you check your progress.

There is also an interactive CD-ROM inside the front cover. This contains more exam-style practice questions, followed by a Unit Test for each of Units 1 and 2. The Unit Tests have been written to be as close as possible to the actual Edexcel Test papers and so will provide you with a true benchmark of the standard that you have reached.

Between them, the book and the CD-ROM should provide you with all the guidance and information needed to enable you to face your exams with confidence.

Unit 1

An elderly physicist was asked how much he had in the bank. 'How much of what?' he responded.
'Money, of course!'
'Fifteen million, three hundred thousand, one hundred and four,' he replied.
The physicist was not a rich man. He had quoted his balance in Turkish Lira, which, at that time, had an exchange rate of 2.6 million to the pound (1.4 million to the American dollar).
The story has relevance to measurements in physics. It is meaningless to state that the size of a wire is 10; we must state the **quantity** that is measured (e.g. the length of the wire) and the **unit** (e.g. cm).
In this section, and throughout this book, you will identify and use a number of base quantities (and their units) that are fundamental to all physical measurements. You will develop and use derived units for quantities for a wide range of physical properties.

Tip

Many students lose marks in examinations by failing to include the unit of a derived quantity! Always show the unit for all calculated quantities.

1.1 Physical quantities, base and derived units

All measurements taken in physics are described as physical quantities. There are seven quantities fundamental to physics. These are mass, length, time, temperature, current, amount of substance and luminous intensity. All other quantities are derived from these **base** quantities – for example, speed is distance (length) divided by time.

SI units

A system of measurement is needed so that a comparison of the sizes can be made with other people's values. Over the years, many different systems of units have been used: in the UK and US, pounds and ounces, degrees Fahrenheit and miles are still common measurements of mass, temperature and length. Scientists have devised an international system that uses agreed **base units** for the seven base quantities. These are termed **SI units** (abbreviated from the French Système International d'Unités).

The base units needed for AS (and A2) examinations are defined in Table 1.1.

Base quantity	Base unit	Symbol	Definition
Length	Metre	m	The distance travelled by electromagnetic radiation through a vacuum in a time of $\frac{1}{299\,792\,458}$ second
Mass	Kilogram	kg	The mass of a standard platinum–iridium cylinder held in Sèvres, France
Time	Second	s	9 192 631 770 periods of the radiation emitted from an excited caesium-133 atom
Current	Ampere	A	The current that, when flowing in two infinitely long parallel wires placed one metre apart in a vacuum, produces a force per unit length of $2 \times 10^{-7}\,\text{N}\,\text{m}^{-1}$
Temperature interval	Kelvin	K	$\frac{1}{273.16}$ of the thermodynamic temperature difference between absolute zero and the triple point of water
Amount of substance	Mole	mol	The amount of substance that contains the same number of elementary particles as there are atoms in 12 grams of carbon-12

Table 1.1 ▶
Base quantities and units

Derived units

Many physical quantities are defined in terms of two or more base quantities. Some examples you may be familiar with are area, which is the product of two lengths, speed, which is the distance (length) divided by time, and density, which equals mass divided by volume (length cubed).

Table 1.2 includes some of the derived quantities and units that will be used in this book. Fuller definitions will be given in later chapters.

Quantity	Definition	Derived unit	Base units
Speed	$\dfrac{distance}{time}$		$m\,s^{-1}$
Acceleration	$\dfrac{\Delta velocity}{time}$		$m\,s^{-2}$
Force	$F = m \times a$	newton (N)	$kg\,m\,s^{-2}$
Pressure	$p = \dfrac{F}{A}$	pascal (Pa)	$kg\,m^{-1}\,s^{-2}$
Work (energy)	$W = F \times x$	joule (J)	$kg\,m^2\,s^{-2}$
Power	$P = \dfrac{W}{t}$	watt (W)	$kg\,m^2\,s^{-3}$
Charge	$Q = I \times t$	coulomb (C)	$A\,s$
Potential difference	$V = \dfrac{W}{Q}$	volt (V)	$kg\,m^2\,A^{-1}\,s^{-3}$
Resistance	$R = \dfrac{V}{I}$	ohm (Ω)	$kg\,m^2\,A^{-2}\,s^{-3}$

Table 1.2 ◄
Derived units

Worked example

Use the definitions from Table 1.2 to show that the pascal can be represented in base units as $kg\,m^{-1}\,s^{-2}$.

Answer

$$\text{Force (N)} = \text{mass (kg)} \times \text{acceleration } (m\,s^{-2})$$

$$\text{Pressure (Pa)} = \frac{\text{force } (kg\,m\,s^{-2})}{\text{area } (m^2)}$$

$$\Rightarrow \text{ the unit Pa} = \frac{kg\,m\,s^{-2}}{m^2}$$

$$= kg\,m^{-1}\,s^{-2}$$

Prefixes

Many measurements are very much larger or smaller than the SI base unit. The thickness of a human hair may be a few millionths of a metre and the voltage across an X-ray tube hundreds of thousands of volts. It is often useful to write these as multiples or sub-multiples of the base unit. You will probably be familiar with the kilometre (1 km = 1000 m) and the millimetre (1 mm = 0.001 m). Table 1.3 gives the prefixes commonly used in the AS (and A2) courses.

Prefix	Symbol	Multiple	Example
pico	p	10^{-12}	$1\,pF = 10^{-12}\,F$
nano	n	10^{-9}	$1\,nA = 10^{-9}\,A$
micro	µ	10^{-6}	$1\,\mu V = 10^{-6}\,V$
milli	m	10^{-3}	$1\,mm = 10^{-3}\,m$
kilo	k	10^{3}	$1\,kW = 10^{3}\,W$
mega	M	10^{6}	$1\,M\Omega = 10^{6}\,\Omega$
giga	G	10^{9}	$1\,GHz = 10^{9}\,Hz$
tera	T	10^{12}	$1\,Tm = 10^{12}\,m$

Table 1.3 ▲
Standard prefixes

Worked examples

1 A metal sphere has a radius of 3.0 mm and a mass of 0.96 g. Calculate the volume of the sphere and determine the density of the metal.

2 The resistance of a conductor is given by the equation

$$R = \frac{V}{I}$$

Calculate the potential difference, V, needed for a current, I, of 25 µA to flow through a 2.2 kΩ resistor. Give your answer in mV.

Answers

1 Volume $= \frac{4}{3}\pi r^3 = \frac{4}{3}\pi (3.0 \times 10^{-3}\,\text{m})^3 = 1.1 \times 10^{-7}\,\text{m}^3$

Density $= \dfrac{\text{mass}}{\text{volume}} = \dfrac{9.6 \times 10^{-4}\,\text{kg}}{1.1 \times 10^{-7}\,\text{m}^3} = 8.5 \times 10^3\,\text{kg}\,\text{m}^{-3}$

2 Rearrange the equation and write the quantities in standard form.

$$V = IR = 25 \times 10^{-6}\,\text{A} \times 2.2 \times 10^3\,\Omega = 0.055\,\text{V} = 55\,\text{mV}$$

1.2 Scalar and vector quantities

If you travel from town A to town B, it is unlikely that you will follow a direct route between the towns (see Figure 1.1).

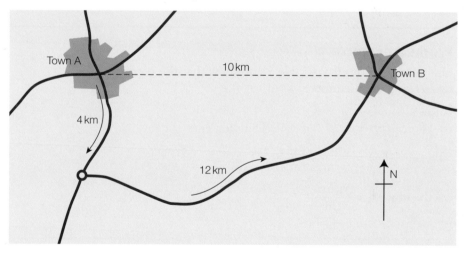

Figure 1.1 ▶
Distance and displacement

Assume that town B is 10 km due east of A and that the journey by road is 16 km. On arrival at B, you will have travelled a **distance** of 16 km and you will have been **displaced** from your starting point by 10 km **due east**. Distance gives only a magnitude (size) of the journey's length, while displacement needs both magnitude and direction.

Quantities represented solely by their magnitude are called **scalars**, and those with both magnitude and direction are **vectors**.

If the journey takes two hours, the average **speed** $\left(\frac{\text{distance}}{\text{time}}\right)$ is 8 km h^{-1}, whereas the average **velocity** $\left(\frac{\text{displacement}}{\text{time}}\right)$ is 5 km h^{-1} due east.

When adding scalar quantities, normal arithmetic is applied. In the example above, the total distance travelled after returning to town A will be 16 km + 16 km, which equals 32 km. However, you would be back at your starting point, so your displacement (10 km due east plus 10 km due west) is zero.

Worked example

An aeroplane flies a distance of 1150 km from London to Oslo in 1 hour and 30 minutes. The return journey follows a different flight path and covers a distance of 1300 km in 2 hours. Calculate:

1 the average speed of the aircraft

2 its average velocity over the two trips.

Answer

1 $\quad$ Average speed $= \dfrac{\text{distance}}{\text{time}} = \dfrac{2450\,\text{km}}{3.5\,\text{h}}$

$\qquad\qquad\quad = 700\,\text{km}\,\text{h}^{-1}$

2 $\quad$ Average velocity $= \dfrac{\text{displacement}}{\text{time}} = \dfrac{0\,\text{km}}{3.5\,\text{h}}$

$\qquad\qquad\quad\ = 0\,\text{km}\,\text{h}^{-1}$

Tip

Throughout this book, quantity algebra is used in all calculations. This means that when the values of quantities are substituted into an equation the appropriate unit is also included. In later examples, where several quantities are given as a multiple or sub-multiple of the SI unit, it can be useful in ensuring that the final answer has the correct magnitude. Although this is a recommended practice, it is not a requirement in the AS and A2 examinations.

Addition of vectors requires direction to be considered. It can be achieved with the help of scale drawings.

If, after travelling to town B, you move 10 km due north to town C, your resultant displacement will be 14 km at an angle of 45° to AB (see Figure 1.2).

Worked example

An oarsman rows a boat across a river with a velocity of 4.0 m s^{-1} at right angles to the bank. The river flows parallel to the bank at 3.0 m s^{-1}. Draw a scale diagram to determine the resultant velocity of the boat.

Answer

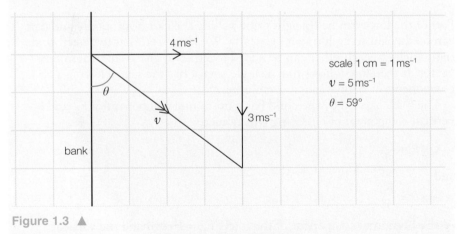

scale 1 cm = 1 ms^{-1}

$v = 5\,\text{ms}^{-1}$

$\theta = 59°$

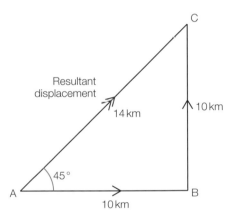

Figure 1.2 ▲
Vector addition

Figure 1.3 ▲

The resultant velocity is 5 m s^{-1} at 59° to the bank.

For the addition of two vectors at right angles, the laws of Pythagoras and trigonometry can be used to determine the resultant vector.

In the above example, the magnitude of the resultant velocity, v, is given by the expression:

$\quad v^2 = (4.0\,\text{m s}^{-1})^2 + (3.0\,\text{m s}^{-1})^2$

$\Rightarrow v = 5.0\,\text{m s}^{-1}$

and the direction is found using the expression:

$\quad \tan\theta = \dfrac{4.0\,\text{m s}^{-1}}{3.0\,\text{m s}^{-1}}$

$\Rightarrow \theta = 59°$

These methods will be used in Chapter 4 to add forces and in the A2 course to combine magnetic fields.

Table 1.4 gives a list of some scalar and vector quantities.

Scalar	Vector
Distance	Displacement
Speed	Velocity
Time	Acceleration
Mass	Force
Amount of substance	Weight
Temperature	Momentum
Charge	Magnetic flux density
Energy	Electric field strength
Potential difference	
Resistance	

Table 1.4 ▶
Scalars and vectors.

1.3 Vector addition and the resolution of vectors

Imagine a child is pulling a toy cart. The child exerts a force at an angle, θ, to the ground, as shown in Figure 1.4.

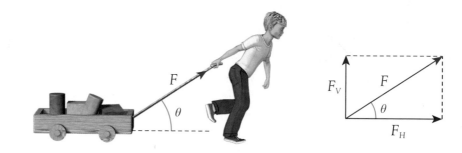

Figure 1.4 ▶

Part of the force can be thought to be pulling the cart horizontally and part can be thought to be lifting it vertically. We have seen that two vectors at right angles to each other can be added together to give a single resultant vector. It therefore follows that a single vector has the same effect as two **components** at right angles to each other.

The force exerted by the child has a horizontal component of F_H and a vertical component of F_V. Using trigonometry:

$$\cos\theta = \frac{F_H}{F}$$

and

$$\sin\theta = \frac{F_V}{F}$$

$\Rightarrow$ the horizontal component of the force, $F_H = F\cos\theta$ and the vertical component, $F_V = F\sin\theta$.

Tip

The component adjacent to θ is always $F\cos\theta$. You may find it easier to use the cosine function for both components – for example, if $\theta = 30°$, the horizontal component will be $F\cos 30$ and the vertical component $F\cos 60$.

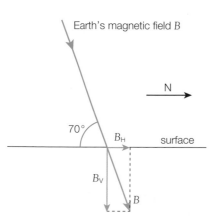

Figure 1.5 ▲

Worked example

The Earth's magnetic field has a flux density of about $5.0 \times 10^{-5}\,\text{T}$ in the UK and enters the Earth at about $70°$ to the surface. Calculate the horizontal and vertical components of the field.

Answer

$$\text{Horizontal component} = (5.0 \times 10^{-5}\,\text{T}) \times \cos 70°$$
$$= 2.3 \times 10^{-5}\,\text{T}$$
$$\text{Vertical component} = (5.0 \times 10^{-5}\,\text{T}) \times \sin 70°$$
$$= 4.5 \times 10^{-5}\,\text{T}$$

The study of projectiles in Section 3.2 is simplified by taking horizontal and vertical components of the initial velocity of the projectile. The vertical and horizontal motions can then be treated separately.

REVIEW QUESTIONS

1 a) Which of the following is a base quantity?

   A Length   B Metre

   C Second   D Speed

  b) Which of the following is a derived unit?

   A Force   B Metre

   C Newton   D Second

  c) Which of the following is a scalar quantity?

   A Acceleration   B Force

   C Metre   D Speed

  d) Which of the following is a vector quantity?

   A Distance   B Mass

   C Time   D Velocity

2 Calculate the force applied to a surface of area 220 mm²
  to create a pressure of 10 MPa. Give your answer in kN.

3 The capacitance, C, of a capacitor is defined by the
  equation:

$$C = \frac{Q}{V}$$

  where Q is the charge stored and V is the potential
  difference across the capacitor.

  Use Table 1.2 to show that the unit of capacitance –
  the farad, F – may be represented as $kg^{-1} m^{-2} A^2 s^4$ in
  base units.

4 A hiker walks 5 km due east and then 10 km due south.

  a) Draw a scale diagram to calculate the displacement
     of the hiker from the starting position.

  b) Calculate the average speed and the average
     velocity of the hiker if the complete journey took
     5 hours.

5 An aeroplane takes off with a velocity of $120\,m\,s^{-1}$ at
  an angle of 30° to the runway. Calculate the horizontal
  and vertical components of the velocity.

2 A guide to practical work

Physics is a very practical subject, and experimental work should form a significant part of your AS Physics course. Indeed, it is easier to learn and remember things if you have actually **done** them rather than having read about them or been told about them. That is why this book is illustrated throughout by experiments, usually with a set of data for you to work through. Nothing is like the 'real thing', however, so you should be carrying out many of these experiments for yourself in the laboratory. You will need to learn certain experiments mentioned in the Edexcel specification, and you may be asked to describe these in the examination. Other experiments are designed to illustrate and give you a better understanding of the theory. Some are particularly designed to develop your practical skills, which will be tested as part of Unit 3 – the internal assessment. As part of the internal assessment, you will be tested on:
- planning
- implementation and measurements
- analysis.

Before we look at these aspects of practical work, however, we need to be clear about certain terminology used in physics, particularly precision, accuracy and errors.

2.1 Errors: accuracy and precision

In everyday English, accuracy and precision have similar meanings, but in physics this is not the case. Their meanings are not the same and the difference must be understood. This is best illustrated by considering an archer shooting arrows at a target. In this situation, precision means getting all the shots close together and accuracy means getting them where they should be (the bull's eye!).

Figure 2.1 ▶
Precision and accuracy

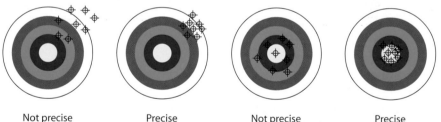

| Not precise
Not accurate | Precise
Not accurate | Not precise
Accurate | Precise
Accurate |

A good archer may achieve high **precision**, but if the sights of the bow are not adjusted properly, the **accuracy** will be poor. This gives rise to a **systematic error**. A poor archer with correctly adjusted sights will scatter the arrows around the bull or, in terms of physics, will make **random errors**.

  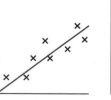

| Random and
systematic error | Systematic error | Random error | Little random or
systematic error |

Figure 2.2 ▲
Random and systematic errors

In physics experiments:

- **systematic errors** can be minimised by taking sensible precautions, such as checking for zero errors and avoiding parallax errors, and by drawing a suitable graph
- **random errors** can be minimised by taking the average of a number of repeat measurements and by drawing a graph that, in effect, averages a range of values.

Consider a mechanical micrometer screw gauge. Such an instrument has high **precision** as it can read to a precision of 0.01 mm. However, its **accuracy** will depend on how uniformly the pitch of the screw has been manufactured and whether or not there is a zero error. You cannot do much about the former, although misuse, such as overtightening, can damage the thread. You should **always** check for zero error before using any instrument and make allowance for it if any is present. This is particularly true for digital instruments, which we often take for granted will be accurate and thus do not check for any zero error. You should also understand that the accuracy of digital instruments is determined by the quality of the electronics used by the manufacturer.

Exercise

A student investigates the motion of a tennis ball rolling down a slope between two metre rules using the arrangement shown in Figure 2.3.

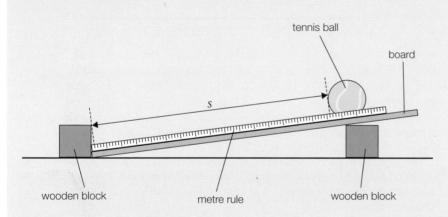

tennis ball

board

s

wooden block

metre rule

wooden block

Figure 2.3 ▲ Random and systematic errors

The student uses a digital stopwatch to find the time, t, that it takes for the ball to roll a distance, s, down the slope. She devises a technique in which she releases the ball from different markings on the metre rule and times how long it takes for the ball to roll down the slope and hit the block at the bottom. She records the results as shown in Table 2.1.

s/m	0.200	0.400	0.600	0.800	1.000
t/s	0.70	1.11	1.35	1.57	1.77
t^2/s^2					

Table 2.1 ▲ Results

The student thinks that s and t are related by the equation of motion:

$$s = \frac{1}{2}at^2$$

where a is the acceleration of the ball down the slope, which she assumes to be uniform. To test her hypothesis, she decides to plot a graph of t^2 against s.

1 Complete the table by adding values for t^2 and then plot a graph of t^2 against s.

2 Discuss the extent to which the graph confirms the student's hypothesis.

Note

You need to be able to rearrange an equation into the linear form

$$y = mx + c$$

For the proposed relationship $s = \frac{1}{2}at^2$, you need to realise that s and t are the variables and a is a constant.

Then a graph of s on the y axis against t^2 on the x axis should produce a linear graph of gradient $\frac{1}{2}a$. In this case, $c = 0$, so the graph goes throught the origin.

It is conventional to plot the *controlled* variable (s in this example) on the x axis and so this is why the student decides to plot a graph of t^2 on the y axis against s on the x axis rather than the other way around.

3 Determine the gradient of the graph and hence calculate a value for the acceleration of the ball down the slope.

4 Explain how the graph reduces the effects of random and systematic error in the experiment.

5 Suggest the possible causes of these errors.

Answer

1 You should obtain a graph like that shown in Figure 2.4. Note that a large but convenient scale has been chosen for each axis and that the axes have been labelled with a forward slash between the physical quantity being plotted and its unit. This is a convention that you should adopt.

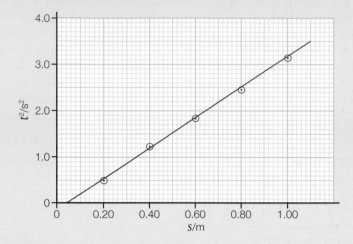

Figure 2.4 ▲

2 The equation of the line should be:

$$t^2 = \frac{2s}{a}$$

If this is correct, the graph should be a straight line of gradient $\frac{2}{a}$ passing through the origin. The graph **is** a straight line, but it does **not** go through the origin. This could be because there is an additional constant term in the equation or because there may be a small systematic error in the readings. The latter would suggest a value for s of about 0.04 m (4 cm) when $t = 0$ (see Question 5 below).

3 Using a **large triangle** and remembering that the graph does not pass through the origin, you should find the gradient to be about 3.25 ($s^2 m^{-1}$).

The gradient is equal to $\frac{2}{a}$, so the acceleration is $\dfrac{2}{\text{gradient}} = 0.62\,m\,s^{-2}$.

4 The graph reduces the effect of **random** errors by using the straight line of best fit through the points to average out the five values. The gradient will still be equal to $\frac{2}{a}$ even if there is a **systematic** error, so drawing a graph both indicates whether there **is** a systematic error and enables allowances to be made for it.

5 The scatter of the points suggests a **random** error, particularly for the smaller values. This is probably due to timing errors, which are more apparent when the times are short (e.g. only about 1 s for the first two values).

The intercept of about 4 cm when $t = 0$ is caused by a **systematic** error by the student. She wrongly assumes that if she releases the bottom of the ball from, for example, the 20 cm mark on the rule, it will travel 20 cm before hitting the block. In reality, the ball travels about 4 cm less that this – see Figure 2.6 on page 12.

2.2 Planning

You will be expected to:

- identify the variables in an equation and determine the most appropriate instrument to measure them
- identify and take appropriate steps to control any other relevant variables to make it a fair test.

In the tennis ball experiment, the student identified the distance s and the time t as the two variables and measured them appropriately with a metre rule and digital stopwatch, respectively. She identified the acceleration a as being constant. To control this, she should ensure that the angle of the slope is constant – for example, by making sure that the runway does not 'bow' in the middle.

You will also be expected to:

- demonstrate knowledge of correct measuring techniques
- comment on whether repeat readings are appropriate.

Unfortunately, in the tennis ball experiment, the student neither measured the distances correctly nor took repeat readings. In an experiment such as this, several repeat readings are essential as the times are very short and subject to human error in starting and stopping the stopwatch. The student should also have gone back and taken timings for extra values of s, as five points are not really enough for a graph.

In general, measurements should be repeated, if practically possible, to provide a check against a misreading and to allow reductions in random errors by averaging two or more values. For example, when finding the diameter of a length of wire, the micrometer (or digital callipers) should first be checked for zero error. The diameter should then be measured at each end of the wire and at the centre (to check for taper), taking readings at right angles to each other at each point (to check for uniformity of cross section). This is shown in Figure 2.5.

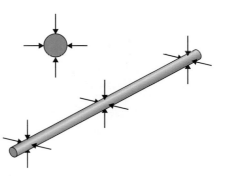

Figure 2.5 ▲

In some experiments, such as finding the current–voltage characteristics of a filament lamp, it would be completely **wrong** to take repeat readings, as the lamp will heat up over time and its characteristics will change. In such a situation you should **plan** for this by making sure that you take a sufficient number of readings the first time around.

You will further be expected to:

- discuss how the data collected will be used
- comment on possible sources of uncertainty and/or systematic error.

The student suggested a suitable graph to test the proposed relationship between distance and time in the tennis ball experiment, and we have discussed the results in terms of sources of random and systematic errors.

We sometimes use the word **uncertainty** rather than error, as 'error' technically means a statistical consideration of the likely error, which is beyond the knowledge needed for Physics A level. 'Uncertainty' means the realistic amount by which you consider your reading could be erroneous. For example, in the tennis ball experiment, the uncertainty in the length s is likely to be the precision with which the ball can be placed on the scale reading of the metre rule – probably ±2 mm. The uncertainty in the timing is most definitely **not** the precision of the stopwatch (0.01 s). The uncertainty is governed by the human reaction time in starting and stopping the stopwatch. Although these effects tend to cancel out to some extent, the uncertainty would still be considered to be in the order of 0.1 s.

The reliability of an experiment can be assessed best by expressing the uncertainties as **percentages**. In the tennis ball experiment, the percentage uncertainty in the measurement of the distance s = 0.200 m would be:

$$\frac{2\,\text{mm}}{200\,\text{mm}} \times 100\% = 1\%$$

The percentage uncertainty in the corresponding time, $t = 0.70\,\text{s}$, would be:

$$\frac{0.1\,\text{s}}{0.70\,\text{s}} \times 100\% = 14\%$$

This relatively large percentage uncertainty explains why there is noticeable scatter at the beginning of the graph in Figure 2.4.

Exercise

When the student shows her teacher the results of the tennis ball experiment, he suggests that she develops the investigation further by comparing her value for the acceleration, a, with the theoretical value. He tells her that this is:

$$a = \frac{3}{5}g\sin\theta$$

where θ is the angle between the slope and the horizontal bench top.

1 Show that the angle θ should be about 6°.

2 Estimate the percentage uncertainty in attempting to measure this angle with a protractor and comment on your answer.

3 Draw a diagram to show how you could use a trigonometric method to determine $\sin\theta$ and comment on the advantage of this technique.

Answer

1 If $a = \frac{3}{5}g\sin\theta$

$$\Rightarrow \sin\theta = \frac{5a}{3g} = \frac{5 \times 0.62\,\text{m s}^{-2}}{3 \times 9.8\,\text{m s}^{-2}} = 0.105$$

$$\Rightarrow \theta = 6.05° \approx 6°$$

2 Using a protractor, the angle could probably not be measured to better than an uncertainty of ±1°, which gives a percentage uncertainty of:

$$\frac{1°}{6°} \times 100\% = 17\%$$

This is a large uncertainty, which suggests that using a protractor is not a suitable technique.

3 Figure 2.6 shows how a trigonometric technique could be used. It also shows why the student's measurements for s have a systematic error of about 4 cm.

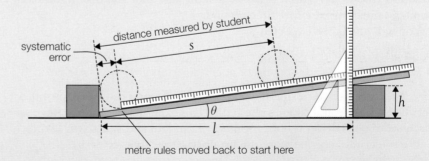

systematic error

distance measured by student

s

θ

h

l

metre rules moved back to start here

Figure 2.6 ▲

In the arrangement shown in Figure 2.6, the distance l was 1.000 m and the height h was 103 mm. These measurements give a value for $\tan\theta$ of:

$$\frac{103\,\text{mm}}{1000\,\text{mm}} = 0.103$$

and hence a value for $\sin\theta$ of 0.102. This has been determined to three significant figures compared with a mere one significant figure using a protractor.

The advantage of this trigonometric method is that it involves two longish lengths, which can each be measured to an uncertainty of ±2 mm or better. This reduces the percentage uncertainty for $\sin\theta$ to about 1% or 2% at most.

Tip

You should always illustrate your work with **diagrams**. These should be carefully drawn, preferably in pencil, using a rule for straight lines. Distances should be marked precisely and any special techniques should be shown – for example, in this case, the use of a set square to ensure that the half-metre rule is vertical.

Implementation and measurements

You will be expected to:

- use the apparatus correctly and apply techniques correctly at all times
- be well organised and methodical, using an appropriately sequenced step-by-step procedure
- record all your measurements using the correct number of significant figures, tabulating measurements where appropriate.

In the example on page 12, using the correct technique shown in Figure 2.6 to determine each value of *s* would enable readings to be taken quickly and accurately and would also allow several repeats of each value to be taken.

Exercise

A student planned to investigate a property of a prism. He wanted to find how the deviation, δ, produced by the prism depends on the angle of incidence, *i*, of a laser beam striking the prism.

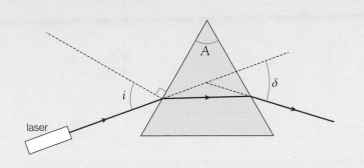

Figure 2.7 ▲

The student placed the prism on a sheet of white paper and drew around it. He removed the prism and drew lines at angles of incidence of 30°, 40°, 50°, 60°, 70° and 80° using a protractor. He then replaced the prism and shone the laser beam along each of the lines in turn, taking particular care not to look directly into the laser beam. For each angle of incidence, he marked the direction of the emerging beam and determined the respective deviation. He recorded the data as shown in Table 2.2.

i/°	30.0	40.0	50.0	60.0	70.0	80.0
δ/°	47.0	39.0	38.0	39.5	42.0	49.5

Table 2.2 ◀
Results

Note that the student took his measurements methodically by initially drawing angles of incidence at 10° intervals and recorded them in a suitable table with units. He attempted to measure the angles to a precision of 0.5° by interpolating between divisions on the protractor. This is reflected in the table by the student giving all of the angles to an appropriate number of significant figures.

1 Plot a graph of the deviation, δ, against the angle of incidence, *i*.

2 Use your graph to determine the angle of minimum deviation, δ_{min}.

3 Suggest extra readings that could be taken to improve the graph.

4 The student looked in a book on optics and found that the refractive index, μ, of the material of the prism is given by the equation:

$$\mu = \frac{\sin \dfrac{A + \delta_{min}}{2}}{\sin \dfrac{A}{2}}$$

where *A* is the angle of the prism. The student measured *A* and found it to be 60.0°. Use this information to find a value for the refractive index.

Answer

1 Your graph should look like that in Figure 2.8.

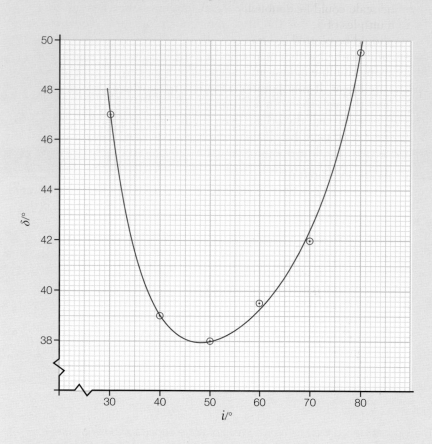

Figure 2.8 ▲

2 The minimum deviation is 37.9°.

3 The angle of minimum deviation could be determined with more certainty if extra readings were taken for $i = 45°$ and $i = 55°$.

4

$$\mu = \frac{\sin \dfrac{A + \delta_{min}}{2}}{\sin \dfrac{A}{2}} = \frac{\sin \dfrac{60.0 + 37.7}{2}}{\sin \dfrac{60}{2}} = \frac{\sin 49}{\sin 30} = 1.51$$

You will also be expected to:

● obtain an appropriate number of measurements
● obtain an appropriate range of measurements.

This is achieved by taking readings over the widest possible range (readings with i less than about 40° are not achievable as total internal reflection occurs) and by subsequently taking extra readings at $i = 45°$ and $i = 55°$.

2.3 Analysis

You will be expected to:

● produce a graph with sensible scales and with appropriately labelled axes, including correct units
● plot points correctly
● draw the straight line, or smooth curve, of best fit.

The graph in Figure 2.8 is an example of when the most suitable scale can be achieved by **not** starting the axes at the origin. You should always choose a scale that occupies at least half the graph paper in both the *x* and *y* directions (or else the scale could be doubled!) and that avoids awkward scales such as scales of multiples of 3.

Points should be plotted accurately, with interpolation between the scale divisions, and marked with a neat cross or a small dot with a ring round it. When drawing the line of best fit, you should remember that not all functions in physics are linear. If your points clearly lie on a curve, you must draw a smooth curve through them. In addition, not every straight line will necessarily pass through the origin, even though you may be expecting it to (as in Figure 2.4 on page 10). If this is the case, you need to discuss possible reasons as to why the line does not pass through the origin.

(as in Figure 2.4 on page 10)

You will also be expected to:

- comment on the trend or pattern obtained
- derive the relation between two variables and determine any constant
- discuss and use related physics principles.

In the tennis ball experiment on page 9, we expected to get a straight line through the origin, but we found that there was a small intercept. We then discussed the possible reasons for this and came to the conclusion that a systematic error had arisen as the distance *s* had not been measured correctly.

We established that the relationship between the variables *s* and *t* was $t^2 = \dfrac{2s}{a}$ and that the constant term $\dfrac{2}{a}$ was equal to the gradient of the graph.

You will further be expected to:

- attempt to qualitatively consider sources of error and, where possible, discuss errors quantitatively
- suggest realistic modifications to reduce error or improve the experiment
- provide a final conclusion.

In the tennis ball experiment, we considered, quantitatively, the error likely to occur if a protractor was used to measure the angle of slope. We concluded that an uncertainty of around 17% was unacceptable, so we improved the experiment by using a trigonometric method.

For this experiment, we can conclude from the measurement of the angle of the slope that:

$$\tfrac{3}{5}\,g\sin\theta = \tfrac{3}{5} \times 9.81\,\mathrm{m\,s^{-2}} \times 0.102 = 0.600\,\mathrm{m\,s^{-2}}$$

The experimental value found for the acceleration was $0.62\,\mathrm{m\,s^{-2}}$. The two values differ by:

$$\frac{(0.62 - 0.60)\,\mathrm{m\,s^{-2}}}{0.61\,\mathrm{m\,s^{-2}}} \times 100\% \approx 3\%$$

This is very acceptable experimental error and so we can conclude that, within experimental error, the acceleration of a tennis ball rolling down a slope is given by the formula $a = \tfrac{3}{5}\,g\sin\theta$.

Note in the above experiment that the two values for the acceleration are both **experimental** values as the value of $\sin\theta$ had to be determined. Therefore the **average** value of the acceleration ($0.61\,\mathrm{m\,s^{-2}}$) was used as the denominator.

Finally, you will be expected to:

- work with due regard for safety at **all** times.

This goes without saying. Apparatus must always be treated with care and respect for your own safety and the safety of others and to prevent damage to instruments and equipment. In the prism experiment on page 13, the student

In the prism experiment on page 13

Tip

First plot your points in pencil. If they look right, with no apparent anomalies, ink them in. Then draw your line in pencil. If you are not happy with your line, you can easily rub it out and have another go without also rubbing out the plotted points.

Tip

Remember:

- the % difference between two experimental values is given by:

% difference =

$\dfrac{\text{difference between the values}}{\text{average of the two values}} \times 100\%$

- the % difference between an experimental value and a stated or known value is given by:

% difference =

$\dfrac{\text{difference between the values}}{\text{stated value}} \times 100\%$

was careful not to look directly into the laser beam. You always need to be aware of the potential dangers of the apparatus with which you are working and act accordingly. In Unit 2 in particular, you will be doing a number of electrical experiments. Electricity should always be treated with the greatest respect – even at the low voltages with which you will be working.

REVIEW QUESTIONS

In Questions 1–3, you will be asked to make some simple measurements for yourself, using things that you can probably find at home. If you cannot do this, you can get the data from the first part of each answer and then work through the rest of the question.

1 You will need:
- unopened 250 g packet of butter (or margarine)
- mm scale (e.g. 30 cm rule)

a) Take such measurements as are necessary to determine a value for the density of the 250 g pack of butter.

b) Estimate the percentage uncertainty in each of your measurements and hence discuss whether your value for the density would allow you to decide whether butter would float in water.

2 You will need:
- an unopened packet (500 sheets) of A4-sized 80 gsm printing paper (gsm is the manufacturer's way of writing 'grams per square metre' – that is, $g\,m^{-2}$)
- kitchen scales
- mm scale (e.g. 30 cm rule)

a) i) Use the kitchen scales to find the mass of the packet of paper.

   ii) Hence determine the mass of a single sheet of this A4 paper.

b) i) Take such measurements as are necessary to determine the area of a single sheet of this A4 paper.

   ii) Calculate an experimental value for the 'gsm' of the paper.

   iii) Calculate the percentage difference between your experimental value and the value given by the manufacturer. Comment on your answer.

c) i) Estimate the thickness of a single sheet of paper and hence determine a value for the density of the paper.

   ii) Describe how you could check the thickness of a single sheet of paper using a micrometer screw gauge or a digital callipers.

3 You will need:
- ten 1 p coins (all dated 1993 onwards)
- mm scale (e.g. 30 cm rule)

(If you do not live in the UK, you should use a small coin of your own currency. You will probably be able to find its mass on the internet.)

a) Take such measurements as are necessary to determine the average diameter, d, and thickness, t, of a 1 p coin.

b) Calculate the volume, V, of a coin and hence the density of the material from which the coin is made given that the mass of a 1 p coin is 3.56 g.

c) After 1992, 1 p coins were made of copper-plated mild steel of density $7.8\,g\,cm^{-3}$.

   i) Determine the percentage difference between your value for the density and the value stated for mild steel.

   ii) Comment on your answer with reference to the uncertainties in your measurements.

d) Until 1992, 1 p coins were made of brass (density $8.5\,g\,cm^{-3}$) and were 1.52 mm thick. Discuss whether your experiment would enable you to detect this difference.

4 In an experiment to find the acceleration of a mass falling freely under gravity, a student used an electronic timer to find the time, t, for a small steel sphere to fall vertically through a distance, h. The student recorded the following results:

h/cm	40	60	80	100	120
t/s	0.30	0.38	0.42	0.47	0.52

The student assumed that the relationship between h and t is $h = \frac{1}{2}gt^2$.

a) Show that a suitable graph to plot would be a graph of t^2 against h and that the gradient of this graph is $\frac{2}{g}$.

b) Tabulate values of h and t^2 and then plot a graph of t^2 against h, starting both scales at the origin.

c) Comment on the graph obtained.

d) Use your graph to determine a value for g.

e) Determine the percentage difference between the value you obtain for g and the accepted value of $9.81\,m\,s^{-2}$.

f) i) Comment on the way in which the student expressed the values of h.

   ii) Suggest how the measurements could have been improved.

3 Rectilinear motion

Rectilinear motion means motion along a straight path. We saw in Topic 1 that average speed is distance divided by time and that velocity is a vector quantity defined as displacement divided by time. As all the motion to be studied in this chapter is rectilinear, it follows that velocities will be used throughout.

3.1 Speed, velocity and acceleration

Average and instantaneous velocity

A sprinter accelerates for 20 metres and then maintains a uniform velocity for the remaining 80 metres of the 100-metre sprint. If the total time taken is 10.0 s, the average velocity, v_{ave}, is found using:

$$v_{ave} = \frac{\text{displacement}}{\text{time}} = \frac{100\,\text{m}}{10.0\,\text{s}} = 10.0\,\text{m s}^{-1}$$

During the first 20 m the runner's velocity is continuously increasing to a maximum of 12.0 m s^{-1}. If the displacement for a small period of time, say 0.01 s, during the acceleration, is 0.06 m, the velocity at this instant will be:

$$v = \frac{0.06\,\text{m}}{0.01\,\text{s}} = 6\,\text{m s}^{-1}$$

The instantaneous velocity is strictly defined as the velocity at an instant, which may be much smaller than 0.01 s, but for practical purposes, measurements of velocity taken over a time that is much shorter than the time of the overall measurements will be regarded as instantaneous.

Average and instantaneous velocities are sometime represented by the equations:

$$v_{ave} = \frac{\Delta v}{\Delta t} \qquad v_{inst} = \frac{\delta v}{\delta t}$$

where Δt represents an interval of time and δt represents a very small interval.

Experiment

Measuring average and instantaneous velocities

A trolley is released at the top of an inclined plane and allowed to run to the bottom of the plane. The distance, Δx, travelled by the trolley down the slope is measured, and the time, Δt, is measured using a stopclock. The average velocity is found by dividing the distance moved down the plane by the time taken.

The interrupter card cuts through the light beam, and this time is electronically recorded. The 'instantaneous' velocity at the bottom of the slope is calculated by dividing the length of the card, δx, by the time, δt, taken to cross the beam.

If another gate is placed near the top of the runway and the card cuts one beam after the other, the initial velocity, u, and the final velocity, v, can be measured.

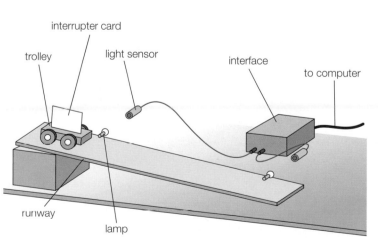

Figure 3.1 ▲
Measuring velocity and acceleration (NB. lamp and sensor supports omitted for clarity)

(Many data-logging interfaces, such as Philip Harris' DL+, will measure the times at each gate and the interval between the gates.)

$$\text{Average velocity} = \frac{\Delta x}{\Delta t} \qquad \text{Initial velocity, } u = \frac{\delta x}{\delta t_1} \qquad \text{Final velocity, } v = \frac{\delta x}{\delta t_2}$$

If the acceleration down the slope is uniform, the average velocity will be

$$\frac{(u + v)}{2}$$

This can be compared with the measured average velocity:

$$\frac{\Delta x}{\Delta t} = \frac{(u + v)}{2}$$

Acceleration

As the trolley moves down the slope, its velocity is steadily increasing. Any change in velocity indicates that the trolley is accelerating. The magnitude of the acceleration is a measure of the rate at which the velocity changes:

$$\text{Acceleration} = \frac{\text{change in velocity}}{\text{time}} = \frac{\Delta v}{\Delta t}$$

The change in velocity is measured in metres per second (m s^{-1}), so acceleration has the unit m s^{-1} per second, which is written as m s^{-2}.

For uniform acceleration, as in the experiment to measure average and instantaneous velocity (page 17), the acceleration is calculated using the equation:

$$a = \frac{(v - u)}{t}$$

Velocity and acceleration are vectors. If an object slows down, $(v - u)$ will be negative. This means that the acceleration is in the opposite direction to the velocities and the object is decelerating.

Worked example

In an experiment, a trolley runs down an inclined plane. An interrupter card of length 20.0 cm cuts through light gates close to the top and bottom of the slope. The following results were recorded from such an investigation:

Time to cut the top gate, $t_1 = 0.30\,\text{s}$

Time to cut the bottom gate, $t_2 = 0.14\,\text{s}$

Time to travel between the gates, $t = 0.50\,\text{s}$

1 Calculate:

a) the velocity of the trolley at each gate

b) the acceleration of the trolley.

Answer

1 Velocity at top gate, $u = \dfrac{(0.200\,\text{m})}{(0.30\,\text{s})} = 0.67\,\text{m s}^{-1}$

Velocity at bottom gate, $v = \dfrac{(0.200\,\text{m})}{(0.14\,\text{s})} = 1.43\,\text{m s}^{-1}$

2 Acceleration, $a = \dfrac{(v - u)}{t} = \dfrac{(1.43\,\text{m s}^{-1} - 0.67\,\text{m s}^{-1})}{0.50\,\text{s}} = 1.5\,\text{m s}^{-2}$

Equations of motion

The motion of an object moving at constant velocity, or accelerating uniformly, can be described by a set of equations known as the equations of motion. The following symbols represent the physical quantities involved in the equations:

s = displacement (m)

u = initial velocity (m s^{-1}) at $t = 0$ s

v = final velocity (m s^{-1})

a = acceleration (m s^{-2})

t = time (s)

The first equation is simply the definition of acceleration rearranged so that the final velocity, v, is the subject of the equation:

$$a = \frac{(v - u)}{t}$$

$$\Rightarrow v = u + at \qquad \text{Equation 1}$$

Average velocity is defined as displacement divided by time. For uniform motion, the average velocity is:

$$\frac{(u + v)}{2}$$

$$\frac{s}{t} = \frac{(u + v)}{2}$$

$$\Rightarrow s = \frac{(u + v)t}{2} \qquad \text{Equation 2}$$

In order that any two quantities can be calculated if the other three are given, a further two equations can be obtained by combining Equations 1 and 2. The resultant expressions are:

$$s = ut + \frac{1}{2}at^2 \qquad \text{Equation 3}$$

$$v^2 = u^2 + 2as \qquad \text{Equation 4}$$

You do not need to be able to perform these combinations, and Equations 3 and 4 will be included on the data sheet at the end of the AS (and A2) examination papers.

Worked example

1 A train starts from rest at a station and accelerates at 0.2 m s^{-2} for one minute until it clears the platform. Calculate the velocity of the train after this time and the length of the platform.

2 The train now accelerates at 0.4 m s^{-2} for the next 540 m. Calculate its final velocity and the time taken to travel this distance.

Answer

1 $u = 0$ m s^{-1} $a = 0.2$ m s^{-2} $t = 60$ s $v = ?$ $s = ?$

Using Equation 1:

$$v = u + at$$

$$= 0 + 0.2 \text{ m s}^{-2} \times 60 \text{ s}$$

$$= 12 \text{ m s}^{-1}$$

Using Equation 3:

$$s = ut + \frac{1}{2}at^2$$

$$= 0 + \frac{1}{2} \times 0.2 \text{ m s}^{-2} \times (60 \text{ s})^2$$

$$= 360 \text{ m}$$

2 $u = 12\,\mathrm{m\,s^{-1}}$ $a = 0.4\,\mathrm{m\,s^{-2}}$ $s = 540\,\mathrm{m}$ $v = ?$ $t = ?$

Using Equation 4:

$$v^2 = u^2 + 2as$$
$$= (12\,\mathrm{m\,s^{-1}})^2 + 2 \times 0.4\,\mathrm{m\,s^{-2}} \times 540\,\mathrm{m}$$
$$= 576\,\mathrm{m^2\,s^{-2}}$$
$$\Rightarrow v = 24\,\mathrm{m\,s^{-1}}$$

Using Equation 1:

$$v = u + at$$
$$24\,\mathrm{m\,s^{-1}} = 12\,\mathrm{m\,s^{-1}} + 0.4\,\mathrm{m\,s^{-2}} \times t$$
$$\Rightarrow t = 30\,\mathrm{s}$$

Tip

Always write down the values of s, u, v, a and t that you are given and then select the appropriate equation to obtain the unknown quantity.

Experiment

Measuring acceleration due to gravity

For an object dropped from rest, the acceleration due to gravity can be calculated using the equations of motion:
$s = ut + \frac{1}{2}at^2$ or $v^2 = u^2 + 2as$.
As $u = 0\,\mathrm{m\,s^{-1}}$ and $a = g$, these equations become
$s(h) = \frac{1}{2}gt^2$ and $v^2 = 2as(h)$.

Method 1

A ball-bearing is held by an electromagnet at height, h, above a trapdoor switch (Figure 3.2a). When the switch is thrown from A to B, the circuit is broken and the ball begins to fall. At that instant the stopclock starts timing. When the ball strikes the trapdoor, the lower circuit is broken and the clock is stopped. The height is measured using a metre rule. The timing is repeated several times, and an average value of t is recorded.

The experiment is repeated for a range of different heights. A graph of h against t^2 is plotted.

As $h = \frac{1}{2}gt^2$, the gradient of the line will be $\frac{1}{2}g$.

Although it would be conventional to plot the controlled variable (h) on the x-axis (see page 9), it is sometimes acceptable to break this convention for convenience, as is the case here.

Method 2

A cylinder is dropped down a plastic tube so that it cuts through a light beam (Figure 3.2b). The length of the cylinder, l, and the height of the top of the tube, h, above the light gate are measured, and the time, t, for the cylinder to cut the beam is recorded.

The experiment is repeated several times, and an average value of t is obtained. The velocity, v, of the cylinder passing through the gate is calculated

using $v = \dfrac{l}{t}$

The experiment is repeated for a range of heights, and a graph of v^2 against h is plotted.

As $v^2 = 2gh$, the gradient of the line will be $2g$.

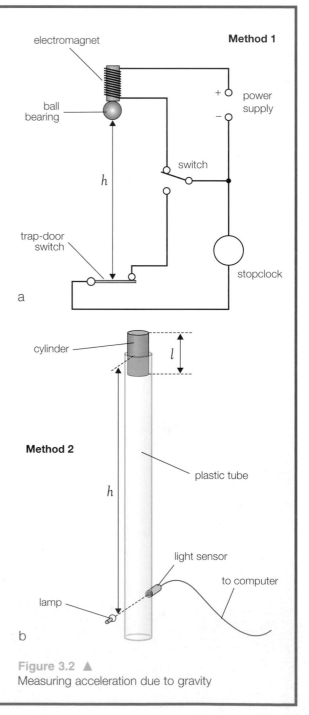

Method 1

Method 2

Figure 3.2 ▲
Measuring acceleration due to gravity

Free fall

In the absence of air resistance, all objects, whatever their mass, will fall freely with the same acceleration. Galileo Galilei tested this hypothesis by dropping different masses from the leaning tower of Pisa. Similarly, the astronauts in the Apollo spacecraft showed that a hammer and a feather fall at the same rate on the Moon.

The acceleration of free fall on Earth, commonly termed the acceleration due to gravity, g, has a value of $9.8\,m\,s^{-2}$.

When using the equations of motion for free-falling bodies, we need to be aware that displacement, velocity and acceleration are vector quantities and that the acceleration due to gravity always acts downward towards the Earth. If an object is thrown upwards, it will still be accelerating downwards at $9.8\,m\,s^{-2}$. If the upward velocity is assigned a positive value, it follows that the value of downward acceleration must be negative. If the body was thrown downwards, its direction would be the same as that of g, so both can be given positive values.

Worked example

A ball is thrown vertically upwards with an initial velocity of $10\,m\,s^{-1}$. Calculate the maximum height it will reach above its starting position and the time it will take to reach this height.

Answer

At the instant the ball is at its maximum height, its velocity will be zero.

$u = +10\,m\,s^{-1}$ $\quad v = 0\,m\,s^{-1}$ $\quad a = -9.8\,m\,s^{-1}$ $\quad s = ?$ $\quad t = ?$

Using $v^2 = u^2 + 2\,a\,s$:

$$0 = (+10\,m\,s^{-1})^2 + 2 \times (-9.8\,m\,s^{-2}) \times s$$
$$100\,m^2s^{-2} = 19.6\,m\,s^{-2} \times s$$
$$s = 5.1\,m$$

Using $v = u + at$:

$$0 = +10\,m\,s^{-1} + (-9.8\,m\,s^{-2}) \times t$$
$$t = 1.0\,s$$

3.2 Projectiles

Objects projected horizontally will still fall freely with an acceleration of $9.8\,m\,s^{-2}$.

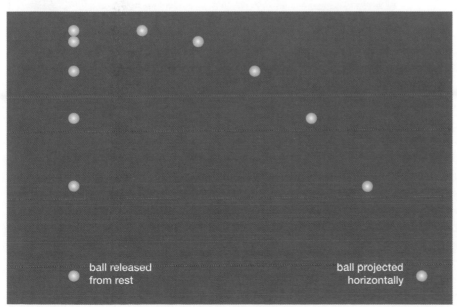

ball released from rest

ball projected horizontally

Figure 3.3 ◄
Falling spheres illuminated by a strobe lamp

Figure 3.3 shows two balls – one released from rest and the other projected horizontally – photographed while illuminated by a strobe lamp. It is clear that the acceleration of the projected ball is unaffected by its horizontal motion.

The horizontal distance travelled by projectiles can be found by considering the vertical and horizontal motions separately. In the vertical plane, the object will accelerate down at $9.8\,\mathrm{m\,s^{-2}}$, while the horizontal velocity remains constant. The equations of motion can be applied to the vertical motion to ascertain the time spent by the object in free fall, and the horizontal displacement is the product of the constant velocity and this time.

Worked example

A tennis ball is volleyed horizontally at a height of 1.5 m at $20\,\mathrm{m\,s^{-1}}$. Calculate the time taken by the ball to hit the court and the horizontal distance travelled by the ball.

Answer
The instant the ball is struck, in addition to moving horizontally it will begin to fall downward due to the gravitational force acting on it. The vertical motion of the ball is identical to that of a ball dropped from rest and falling to the ground.

In the vertical plane: $u = 0\,\mathrm{m\,s^{-1}}$ $\quad a = 9.8\,\mathrm{m\,s^{-2}}$ $\quad s = 1.5\,\mathrm{m}$ $\quad t = ?$

Using $s = ut + \frac{1}{2}at^2$:

$$1.5\,\mathrm{m} = 0 + \frac{1}{2} \times 9.8\,\mathrm{m\,s^{-2}} \times t^2$$

$$t = 0.55\,\mathrm{s}$$

In the horizontal plane: $u = 20\,\mathrm{m\,s^{-1}}$ (constant) $\quad t = 0.55\,\mathrm{s}$

$$s = u \times t$$
$$= 20\,\mathrm{m\,s^{-1}} \times 0.55\,\mathrm{s}$$
$$= 11\,\mathrm{m}$$

Experiment

Monkey and hunter

A monkey hangs from the branch of a tree. The hunter aims his gun accurately at the monkey and fires. The sharp-eyed primate spots the bullet as it leaves the gun, releases its grip on the branch and falls to the ground. Does the monkey survive?

A laboratory model of the situation is shown in Figure 3.4b. The 'gun' is clamped horizontally to the bench so that it is aimed directly at the 'monkey', which is held by the electromagnet two or three metres away. The 'bullet' is fired, breaking the circuit as it leaves the end of the barrel. The 'monkey' is released and falls freely to the ground. Unfortunately, the 'bullet' will accelerate vertically down at the same rate as the 'monkey', and the monkey will be hit if the hunter's initial aim is true.

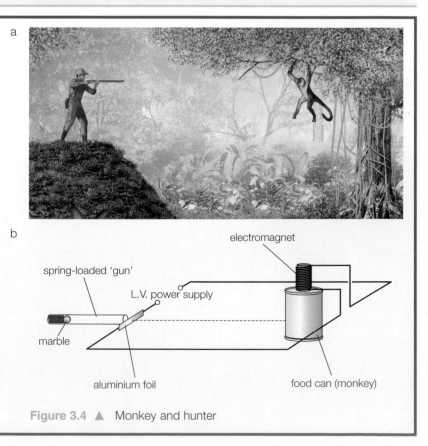

a

b

spring-loaded 'gun'

L.V. power supply

electromagnet

marble

aluminium foil

food can (monkey)

Figure 3.4 ▲ Monkey and hunter

If an object is projected at an angle, the vertical and horizontal motion can still be treated separately by considering the components of the velocity in each plane.

Consider the motion of an object projected at an angle, θ, with an initial velocity, u.

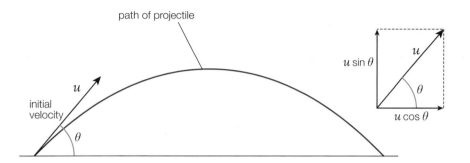

Figure 3.5 ◄
Projectile motion

In the vertical plane, the initial velocity is $u \sin \theta$ (upwards) and the acceleration is $9.8 \, \text{m s}^{-2}$ (downwards). As both are vectors, if a positive value is assigned to the initial velocity, the acceleration, downward velocities and downward displacements will have negative values.

Neglecting air resistance, the horizontal component of the initial velocity will remain constant throughout the motion.

Worked example

A football is kicked with a velocity of $12 \, \text{m s}^{-1}$ at an angle of $30°$ to the ground.

Calculate:

a) the vertical height reached by the ball

b) the time taken for the ball to rise to this height

c) the horizontal displacement of the ball.

Answer

In the vertical plane:

a) $u = 12 \sin 30 \, \text{m s}^{-1} = 6.0 \, \text{m s}^{-1}$ $v = 0 \, \text{m s}^{-1}$ $a = -9.8 \, \text{m s}^{-2}$ $s = h$

Using $v^2 = u^2 + 2as$:

$$(0 \, \text{m s}^{-1})^2 = (60 \, \text{m s}^{-1})^2 + 2 \times (-9.8 \, \text{m s}^{-2}) \times h$$

$$h = \frac{(60 \, \text{m s}^{-1})^2}{(19.6 \, \text{m s}^{-2})}$$

$$= 1.8 \, \text{m}$$

b) $u = 12 \sin 30 \, \text{m s}^{-1}$ $v = 0 \, \text{m s}^{-1}$ $a = -9.8 \, \text{m s}^{-2}$ $t = ?$

Using $v = u + at$:

$$0 \, \text{m s}^{-1} = 12 \sin 30 \, \text{m s}^{-1} + (-9.8 \, \text{m s}^{-2}) \times t$$

$$t = \frac{(6.0 \, \text{m s}^{-1})}{(9.8 \, \text{m s}^{-2})}$$

$$= 0.61 \, \text{s}$$

In the horizontal plane:

c) $u = 12 \cos 30 = 11 \, \text{m s}^{-1}$ $t = 2 \times 0.61 = 1.2 \, \text{s}$

$$s = u \times t$$

$$= 11 \, \text{m s}^{-1} \times 1.2 \, \text{s}$$

$$= 13 \, \text{m}$$

3.3 Displacement–time and velocity–time graphs

Displacement–time graphs

Figure 3.6 shows the displacement–time graph for an object moving at constant velocity.

Uniform velocity is calculated using the equation:

$$\text{Velocity} = \frac{\text{displacement}}{\text{time}} = \frac{\Delta s}{\Delta t}$$

The velocity is represented by the **gradient** of a displacement–time graph.

The motion represented in Figure 3.7 is that of the trolley accelerating down the incline. The gradient of the line gets steeper, which indicates an increase in velocity.

The instantaneous velocity is the gradient, $\frac{\delta s}{\delta t}$, of the graph at a point on the line.

To measure the small values of δs and δt would be very difficult and this would lead to large uncertainties in the measured velocity. The instantaneous velocity is more accurately measured by drawing a tangent to the line at the appropriate point. The gradient of this line is calculated using the larger values Δx and Δt.

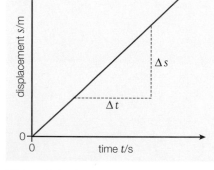

Figure 3.6 ▲
Graph for constant velocity

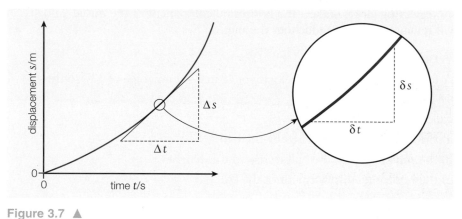

Figure 3.7 ▲
Instantaneous velocity for accelerating object

Experiment

Measuring the displacement of a moving object

The transmitter of a motion sensor sends out pulses of ultrasound and infrared radiation that are picked up by the receiver. The distance of the transmitter from the sensor is continuously recorded. The sensor is interfaced with a computer with data sampling software to measure the position of the trolley at fixed time intervals. The trolley may be pulled along the runway at constant velocity or allowed to accelerate down an inclined plane.

Displacement–time graphs can be drawn from the results or the graphs may be displayed on the computer.

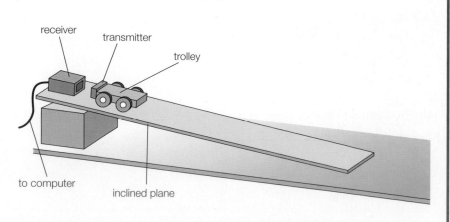

Figure 3.8 ▶

Velocity–time graphs

The results of the displacement–time experiment can be used to illustrate how the velocity of the trolley changes as it moves along the runway. The gradient of the displacement–time graph is taken for a range of times and a velocity–time graph is plotted. This is quite a tricky exercise, and one that is usually better left to the computer program.

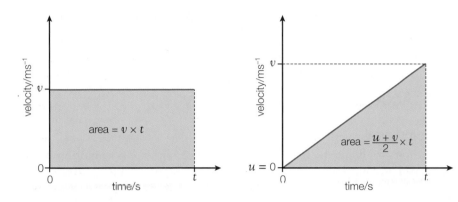

Figure 3.9 ◄
Velocity–time graphs

The graphs in Figure 3.9 represent an object moving at constant velocity and another object with uniform acceleration.

Acceleration is defined by the expression:

$$\text{Acceleration} = \frac{\text{change in velocity}}{\text{time}} = \frac{\Delta v}{\Delta t}$$

The acceleration of an object is therefore equal to the **gradient** of a velocity–time graph. If the object is slowing down, the gradient will be negative. A negative value of acceleration indicates that the vector has the opposite direction to the velocity. This is usually referred to as deceleration.

Finding displacement from velocity–time graphs
Average velocity is displacement divided by time. For uniform motion, this leads to the equation:

$$s = \frac{(u + v)}{2} \times t$$

At constant velocity, the displacement is simply the product of velocity and time and will be the **area** under the horizontal line on the first graph in Figure 3.9.

The area under the line of the accelerating object is $\frac{1}{2} \times$ base $\times$ height of the triangle, which is the same as the expression for displacement given above. (The value of u in Figure 3.9 is zero.)

For any velocity–time graph, the displacement is equal to the area between the line and the time axis.

Worked example

The velocity–time graph in Figure 3.10 represents the motion of a train as it travels from station A to station D.

1 Describe the changes in the motion of the train.

2 Calculate:
   a) the acceleration from A to B
   b) the acceleration from C to D
   c) the total displacement from A to D.

Answer

1 The train accelerates uniformly from A to B, travels at $20\,\text{m s}^{-1}$ until it reaches C and then decelerates, uniformly, to D.

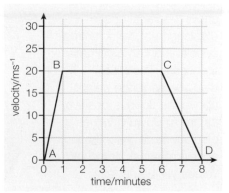

Figure 3.10 ▲
Velocity–time graph

25

2 a) Acceleration = gradient = $\dfrac{20\,\text{m s}^{-1}}{60\,\text{s}}$ = 0.33 m s^{-2}

b) Acceleration = gradient = $\dfrac{-20\,\text{m s}^{-1}}{120\,\text{s}}$ = -0.17 m s^{-2}

c) Displacement

= area under the graph

= $(\frac{1}{2} \times 20\,\text{m s}^{-1} \times 60\,\text{s}) + (20\,\text{m s}^{-1} \times 300\,\text{s}) + (\frac{1}{2} \times 20\,\text{m s}^{-1} \times 120\,\text{s})$

= 7800 m

3.4 Acceleration due to gravity

Bouncing ball

The motion of a bouncing ball provides a good example of how motion is represented graphically.

Experiment

Investigating the motion of a bouncing ball

An advanced motion sensor (e.g. that supplied by ScienceScope) with the appropriate datalogger (ScienceScope's Logbook Timing or Philip Harris' Card logger CL200 or Card logger CL50) is ideal for this investigation. The transmitter is attached to a basketball, or similar large ball, using Blu-tack. The receiver is clamped in position vertically above the transmitter as shown in Figure 3.11. The motion sensor uses a mixture of infrared and ultrasonic signals to measure accurately the distance between the transmitter and the receiver.

The sensor is activated, and the ball is dropped and allowed to bounce two or three times. Practice is needed to ensure that the ball does not rotate and that the transmitter stays beneath the receiver all of the time.

The data can be displayed as a distance–time, velocity–time or acceleration–time graph. The graphs in Figure 3.12 show the variations in displacement, velocity and acceleration on a common timescale. The initial displacement (the height above the floor) is positive, as are all upward values of displacement, velocity and acceleration; all downward values will be negative.

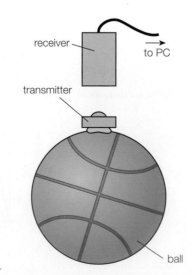

Figure 3.11 ▶

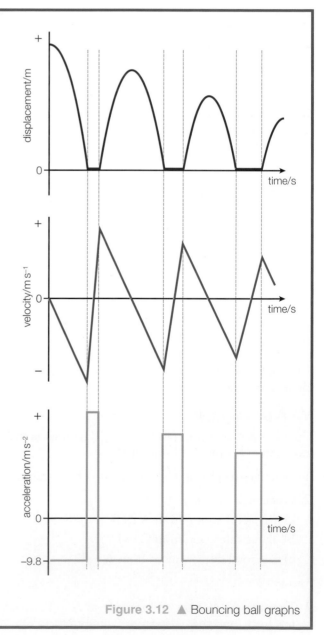

Figure 3.12 ▲ Bouncing ball graphs

The displacement–time graph shows that the initial gradient is zero: the ball starts from rest. The gradient then increases and has a negative value. This means the velocity is downward and is increasing with time. The short time when v is zero is when the ball is in contact with the ground. The gradient becomes positive, showing an upward velocity, and continuously decreases to zero at the top of the bounce. The velocity–time graph has a uniform negative gradient until the ball strikes the ground. This represents the downward acceleration of free fall ($9.8\,\mathrm{m\,s^{-2}}$). Over the contact period, the velocity changes from a large downward value to a slightly smaller upward velocity in a short time. The graph indicates a large, upward acceleration during the deformation and reformation of the ball. The ball accelerates downward as it slows down to zero velocity at the top of the bounce and continues to accelerate as it falls back to the ground.

The acceleration–time chart shows that the ball is accelerating at $-9.8\,\mathrm{m\,s^{-2}}$ whenever it is in the air and that it experiences a large upward acceleration during contact with the ground.

Tip

To describe the motion represented by a graph, always look at the gradient. Check if it is constant, positive, negative or zero. For a displacement–time graph, the gradient will give you the velocity at any time; for a velocity–time graph, the gradient gives you the acceleration.

REVIEW QUESTIONS

1 A train accelerated from rest to a velocity of $40\,m\,s^{-1}$ in a time of 1 minute and 20 seconds.

  a) What was the average acceleration of the train?

    A $0.33\,m\,s^{-2}$ B $0.50\,m\,s^{-2}$

    C $2.0\,m\,s^{-2}$ D $3.0\,m\,s^{-2}$

  b) What was the distance travelled by the train?

    A $48\,m$ B $1600\,m$

    C $3200\,m$ D $4800\,m$

2 What does the gradient of a displacement–time graph represent?

  A Acceleration   B Distance

  C Speed   D Velocity

3 What does the area under a velocity–time graph represent?

  A Acceleration   B Displacement

  C Speed   D Velocity

4 Distinguish between average and instantaneous velocity.

5 a) Define acceleration.

  b) The road-test information for a car states that it can travel from 0–60 mph in 8.0 s.

    i) Estimate the average acceleration of the car during this time.

    ii) Why is the acceleration unlikely to be uniform? ($1\,mph \approx 0.4\,m\,s^{-1}$)

6 Write down the equations of motion.

7 A cyclist travelling at $4.0\,m\,s^{-1}$ accelerates at a uniform rate of $0.4\,m\,s^{-2}$ for 20 s. Calculate:

  a) the final velocity of the cyclist

  b) the distance travelled by the cyclist in this time.

8 a) Describe a method of determining the acceleration due to gravity in the laboratory. Include a labelled diagram of the equipment you would use and describe all the measurements you would take.

  b) State which of your readings is most likely to have the biggest effect on the uncertainty of the final answer and explain why.

9 A stone was dropped down a well. The splash was heard 2.2 s later. Calculate:

  a) the depth of the well

  b) the velocity of the stone when it hit the water.

10 A ball was thrown vertically upward with a velocity of $12\,m\,s^{-1}$ on release. Calculate:

  a) the maximum height from point of release reached by the ball

  b) the time taken for the ball to reach this height

  c) the velocity of the ball 2.0 s after it was released.

11 A football player kicked a ball downfield. The ball left the boot at 30° to the ground and with a velocity of $20\,m\,s^{-1}$. Calculate:

  a) the vertical and horizontal components of the initial velocity of the ball

  b) the time taken for the ball to strike the ground

  c) the horizontal distance travelled by the ball before it bounced.

12

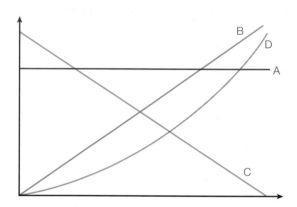

Figure 3.13 ▲

Describe the motion of an object represented by each of the lines A, B, C and D on the graph in Figure 3.13 for:

  a) a displacement–time graph

  b) a velocity–time graph.

13 The graph in Figure 3.14 shows the variation of velocity with time for a body moving in a straight line.

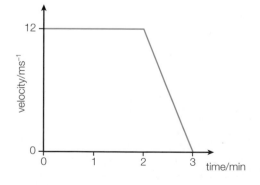

Figure 3.14 ▲

Calculate:

  a) the acceleration of the body during the final minute

  b) the total distance travelled in three minutes

  c) the average velocity over this time.

4 Forces

Forces push or pull and squeeze or stretch. There are several different types of force, most of which will be studied in more detail later. Forces fall into two categories: distant and contact forces.
In this section, you will identify the nature of some forces and draw and interpret free-body force diagrams. You will study the effects of forces on objects using Newton's laws of motion, investigate and apply the expression $\Sigma F = ma$ in situations where m is constant and identify pairs of forces that constitute an interaction between two bodies.

4.1 Nature and types of force

Forces at a distance

Gravitational forces act over very large distances. The planets are kept in orbit around the Sun by the gravitational pull of the Sun. We all experience a gravitational attraction to the Earth: this is called our weight and it pulls us down to Earth if we jump out of an aeroplane or fall off a chair.

The gravitational force between two objects depends on the mass of the objects and their separation. Gravitational forces are very small unless one or both of the masses is extremely large. The attractive force between two elephants standing close together would be very difficult to detect, but the force between a mouse and the Earth is noticeable.

The gravitational force exerted by the Earth on a mass of one kilogram is known as the **gravitational field strength**, g; it has a value of $9.8\,\text{N}\,\text{kg}^{-1}$. It follows that an object of mass, m, will have a weight, W, given by:

$$W = mg$$

Worked example

Estimate the weight of an elephant, a mouse, yourself and a bag of sugar.

Answer
For estimates, g can be taken as $10\,\text{N}\,\text{kg}^{-1}$.

Elephant: mass = 3–7 tonnes, so $W = 5000\,\text{kg} \times 10\,\text{N}\,\text{kg}^{-1} = 50\,\text{kN}$

Mouse: mass 50 grams, so $W = 0.050\,\text{kg} \times 10\,\text{N}\,\text{kg}^{-1} = 0.5\,\text{N}$

Average adult: mass 70 kg, $W = 70\,\text{kg} \times 10\,\text{N}\,\text{kg}^{-1} = 700\,\text{N}$

Sugar: mass 1 kg, $W = 1\,\text{kg} \times 10\,\text{N}\,\text{kg}^{-1} = 10\,\text{N}$

Electrostatic and electromagnetic forces are also examples of forces that act over a distance.

Insulating rods can be given positive or negative charges by rubbing with woollen dusters or cotton rags. The electrostatic forces produced can be investigated by suspending one rod from a thread and holding another rod close to it (Figure 4.1). Unlike gravitational forces, both pulling and pushing effects are observed. Similar charges repel and opposite charges attract. A similar effect is observed when magnets are used.

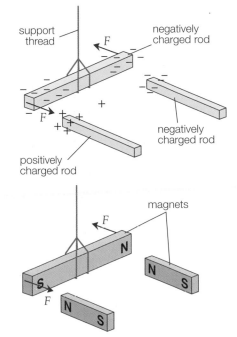

Figure 4.1 ▲
Electrostatic and electromagnetic forces

Contact forces

When you stand on the floor, your weight pushes the floor. The floor and the soles of your shoes are in contact with each other and become slightly compressed. The electrons in the atoms are displaced and short-range forces result. Other examples of contact forces include friction between moving surfaces, viscous forces in liquids and air resistance.

Tension

When a rubber band is stretched, the molecular separation increases. This leads to short-range attractive forces between the molecules. The band is in a state of tension, with the molecular forces trying to restore the band to its original length. All objects subjected to a stretching force are in a state of tension. Examples are the cables of a suspension bridge and tow ropes.

4.2 Forces in equilibrium

Newton's first law of motion

Imagine a stationary rock deep in space, where the gravitational fields of distant stars are negligibly small. The rock experiences no forces and does not move. If the rock was moving it would continue with constant velocity until it felt the gravitational effect of another body in space.

The above ideas seem obvious and trivial nowadays when space missions are commonplace, but they were the bedrock of the laws relating forces and motion that were formulated by Sir Isaac Newton in the seventeenth century. Existing laws were based on earthly experiences – for example, a constant force needs to be applied to a cart to make it continue to move at constant speed in a straight line. Newton was able to envisage a world without the hidden forces of friction, drag and weight.

Newton's first law states that an object will remain in a state of rest or continue to move with a constant velocity unless acted upon by a resultant external force.

Newton's first law explains the dilemma of the moving cart: although a constant force is acting on the cart, it is opposed by an equal set of frictional forces, so there is no **resultant** force acting on it.

Equilibrium

Newton's first law also relates to the 'thought experiment' about the rock in space. If two or more forces were acting on the rock in space, the rock might still remain in a state of rest or uniform motion. Newton's first law adds the provision that no **resultant** force acts on the object. A pair of equal and opposite forces acting on the rock would not affect its motion.

When a number of forces act on a body, and the vector sum of these forces is zero, the body is said to be in **equilibrium**.

Free-body force diagrams

Objects may be subjected to a range of different forces and may themselves exert forces on other bodies.

The rock climber in Figure 4.2 feels his weight and the weight of his rucksack pulling down and the push of the rock on his feet, as well as the pull of the rope on his arms. It is clear that there must be frictional forces between the rope and the hands and the rock face and the climber's boots and that the climber's companion and the rock will experience forces from the climber.

It is very useful to isolate the forces acting on a single object by drawing a **free-body force diagram**. The motion of the climber can be analysed using the free-body force diagram in Figure 4.3. If the climber is in equilibrium, the line of action of all three forces will pass through a single point. In this case, it is simpler to represent the object as a point mass (known as the centre of mass), with all of the forces acting at this point.

Definition

Newton's first law states that an object will remain in a state of rest or continue to move with a constant velocity unless acted upon by a resultant external force.

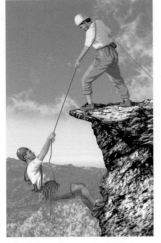

Figure 4.2 ▲
Rock climber

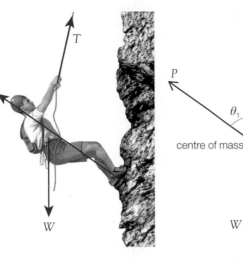

Figure 4.3 ▲
Forces acting on a rock climber

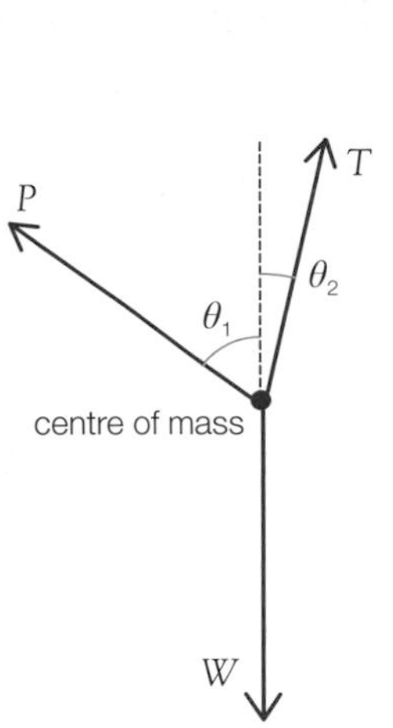

Figure 4.4 ▲
Free-body force diagram of a rock climber

Tip

By representing a body as a point mass it becomes much easier to find the components of the forces.

Experiment

Investigating the equilibrium of three vertical forces

Two pulley wheels are attached to a vertically clamped board. Three weights are connected by thread so that they are in equilibrium, as shown in Figure 4.5.

The direction of the upward forces is found by marking the position of the thread onto a sheet of paper fixed to the board. The angles that forces W_1 and W_2 make with the vertical line drawn between them are measured. For equilibrium:

$$W_1 \cos\theta_1 + W_2 \cos\theta_2 = W_3$$

The experiment is repeated using different values of W_1, W_2 and W_3.

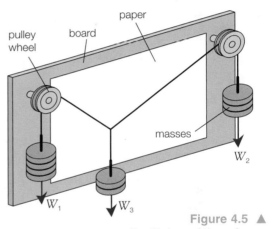

Figure 4.5 ▲
Equilibrium of three forces

An alternative approach is to find the resultant of W_1 and W_2. This is achieved with the help of a scale drawing that uses the parallelogram of forces method of vector addition. The directions of W_1 and W_2 are marked as before, and the lines are drawn to scale to show the magnitudes of W_1 and W_2. A parallelogram is constructed as shown in Figure 4.7.

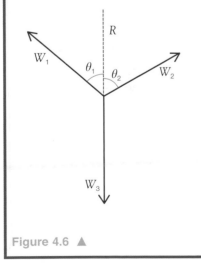

Figure 4.6 ▲

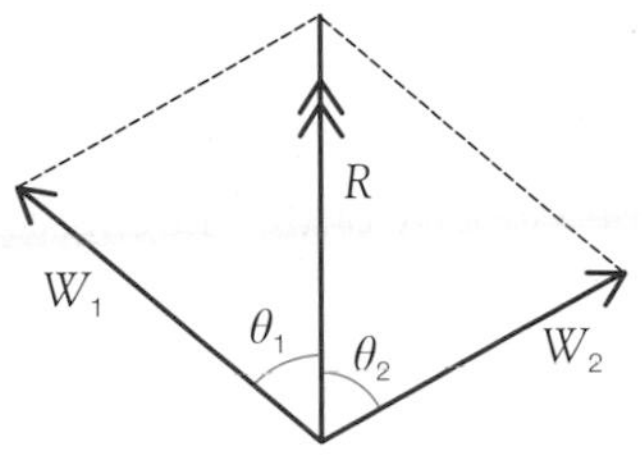

Figure 4.7 ▲ Parallelogram of forces

The parallelogram rule states that the sum of the forces represented in size and direction by adjacent sides of a parallelogram is represented in size and direction by the diagonal of the parallelogram. For equilibrium, the resultant, R, of W_1 and W_2 must be equal in magnitude to W_3 and must be acting vertically upwards.

For equilibrium, the forces can be represented on a scale drawing as a closed triangle that gives a vector sum of zero (see Chapter 1). Alternatively, the sum of the components in the horizontal and vertical planes must both be zero. So, using Figure 4.4:

In the vertical plane:

$(P\cos\theta_1 + T\cos\theta_2) - W = 0$

In the horizontal plane:

$T\sin\theta_2 - P\sin\theta_2 = 0$

Worked example

A 10 kg mass is suspended from a beam using a length of rope. The mass is pulled to one side so that the rope makes an angle of 40° to the vertical, as shown in Figure 4.8.

1 Write expressions for the vertical and horizontal components of the tension, T, in the rope.
2 Use the conditions for equilibrium in the vertical plane to show that the value of T is about 130 N.
3 Use the horizontal component to determine F.

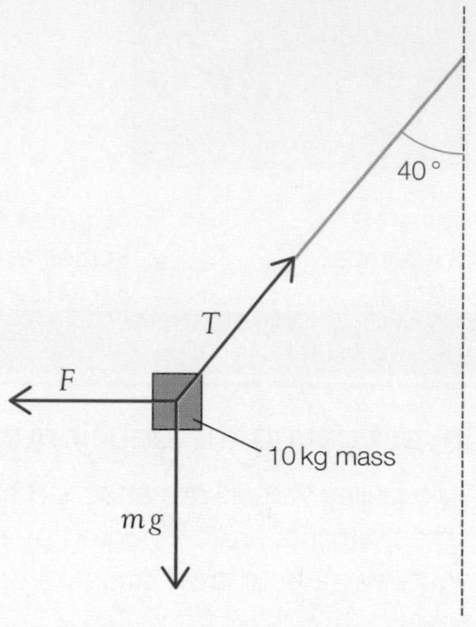

Figure 4.8 ▲

Answer

1 Vertical component of $T = T\cos 40°$
Horizontal component $= T\sin 40°$
2 For equilibrium in the vertical plane:
$T\cos 40° = mg$

$$\Rightarrow T = \frac{10\,\text{kg} \times 9.8\,\text{N kg}^{-1}}{\cos 40°} = 128\,\text{N} \approx 130\,\text{N}$$

3 In the horizontal plane:
$F = T\sin 40° = 128\,\text{N} \sin 40° = 82\,\text{N}$

Tip

For 'Show that…' questions, always calculate the answer to one more significant figure than asked for in the question. For example, in question 2 the value of T is shown as 128 N.

4.3 Newton's second law of motion applied to fixed masses

Definition

Newton's second law states that the acceleration of a body of constant mass is proportional to the resultant force applied to it and in the direction of the resultant force.

Newton's first law states that a body will remain at rest or move with constant velocity if no resultant force acts on it. What will happen to the body if a resultant force does act on the body? Consider the rock in space described on page 30. Give the rock a push and its motion will change. A static rock will move or the velocity of the moving rock will increase, decrease or change direction – that is, it will accelerate. The relationship between the resultant force applied to a body and this acceleration is given in Newton's second law of motion, which states that the acceleration of a body of constant mass is proportional to the resultant force applied to it and in the direction of the resultant force. This definition can be represented by the equation:

$$\Sigma F = ma$$

provided that the unit of force is in newtons (N), the mass is in kilograms (kg) and the acceleration is in metres per second squared (m s^{-2}). It should be noted that this expression applies only for forces acting on fixed masses. This is sufficient for the AS examination, but a more general definition of Newton's second law in terms of momentum change will be dealt with in the A2 book.

Investigating factors that affect the acceleration of an object

The acceleration of the trolley in Figure 4.9 is found by measuring the time taken for the card to cut the light beams and the separation of the gates.

If l is the length of the card, t_1 and t_2 are the times to break the light beams and s is the separation of the gates:

initial velocity, $u = \dfrac{l}{t_1}$

final velocity, $v = \dfrac{l}{t_2}$

The acceleration can be calculated using the equation of motion (see Section 3.1):

$$v^2 = u^2 + 2as$$

The resultant force is provided by the weight on the hanger.

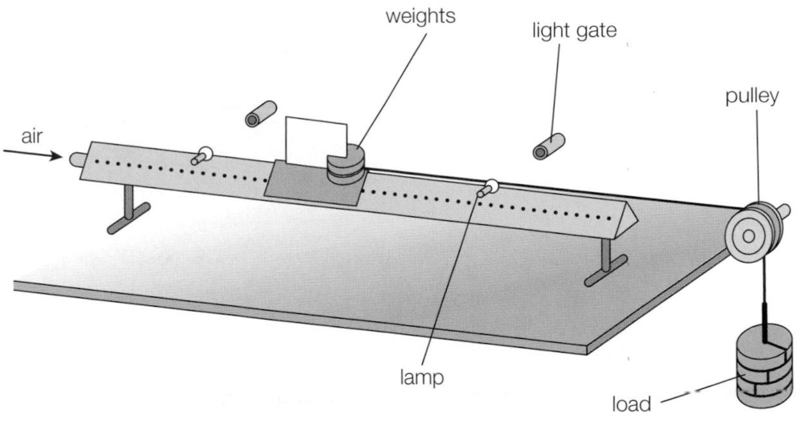

Figure 4.9 ▲
Investigating the acceleration of an object

Effect of the force acting on a fixed mass

It is important to be aware that the gravitational force pulling down on the weights and hanger is acting on the total mass of the system – that is, the mass of the trolley plus the weights and hanger. If extra weights were put onto the hanger, the mass of the system would increase. To make the resultant force increase without changing the mass of the system, four (or more) 10 g weights are initially fixed on the trolley. The force can be increased, without altering the mass, by taking a weight off the trolley and placing it onto the hanger.

A range of forces is used and the corresponding accelerations are measured. A graph of acceleration against force is then plotted.

Effect of mass on the acceleration produced by a fixed force

The weight on the hanger is kept constant. Weights are added to or taken from the trolley to increase or decrease the mass of the system.

The acceleration is found for each mass, as before, and a graph of acceleration against the inverse mass is plotted.

If both graphs are straight lines through the origin, it follows that:

$$a \propto F \quad \text{and} \quad a \propto \frac{1}{m} \quad \Rightarrow a \propto \frac{F}{m}$$

A typical set of results for an experiment finding the effect of the force acting on a fixed mass is as follows:

$l = 0.200\,\text{m} \qquad x = 0.400\,\text{m}$

1 Complete Table 4.1 (use an Excel spreadsheet with appropriate equations in the column headings if you wish) and plot a graph of $a/\text{m s}^{-2}$ against F/N.

2 Use the graph to show that the mass of the trolley is about 0.45 kg.

F/N	t_1/s	t_2/s	u/m s^{-1}	v/m s^{-1}	a/m s^{-2}
0.10	0.50	0.36			
0.20	0.36	0.25			
0.30	0.29	0.20			
0.40	0.25	0.18			
0.50	0.22	0.16			

Table 4.1 ▲

Worked example

The tension in the rope pulling a water skier is 500 N, and the resistive force of the water is 400 N. The total mass of the water skier and skis is 78 kg.

Figure 4.10 ▲ Water skier

1 Draw a free-body force diagram for the water skier.

2 Calculate:

  a) the resultant force

  b) the acceleration of the skier.

Answer

1

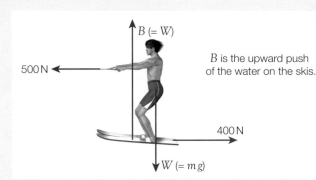

B is the upward push of the water on the skis.

Figure 4.11 ▲ Free-body force diagram for water skier

2 **a)** $F_R = 500\,\text{N} - 400\,\text{N} = 100\,\text{N}$

  b) $a = \dfrac{F}{m} = \dfrac{100\,\text{N}}{78\,\text{kg}} = 1.3\,\text{m s}^{-2}$

Multi-body systems

An engine pulls three carriages along a track. The force of the track on the wheels of the engine drives it forward. If the driving force exceeds the resistive forces, the train will accelerate.

Driving force of tracks on wheels   Total resistive force

Figure 4.12 ▲ Forces on a train

The resultant force is required to accelerate not only the engine but also the carriages:

$$D - F = (m_e + 3\,m_c)\,a$$

The engine and the carriages accelerate at the same rate, but the carriages have no driving force. The free-body force diagram in Figure 4.13 is for a single carriage of the train.

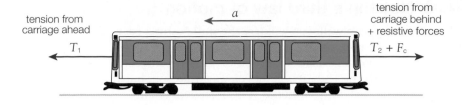

tension from
carriage ahead

a

tension from
carriage behind
+ resistive forces

T_1

$T_2 + F_c$

Figure 4.13 ▲ Free-body force diagram of carriage

The carriage is accelerating like the rest of the train, so it must experience a resultant force. We can apply Newton's second law to the carriage as follows:

$$T_1 - (T_2 + F_c) = m_c a$$

Worked example

A lift and its load are raised by a cable, as shown in Figure 4.14.

1 If the total mass of the lift and its contents is 1.5 tonnes and the tension in the cable is 22.2 kN, calculate the acceleration of the lift.

2 **a)** Draw a free-body force diagram for a woman of mass 50 kg standing in the lift.

b) Calculate the normal contact force of the floor of the lift on her feet.

Answer

1 Resultant force
$= 22\,200\,\text{N} - (1500\,\text{kg} \times 9.8\,\text{m s}^{-2})$
$= 7500\,\text{N}$

$7500\,\text{N} = 1500\,\text{kg} \times a$
$a = 5.0\,\text{m s}^{-2}$ upwards

2 **a)** See Figure 4.15.

b) Resultant force
$= R - (50\,\text{kg} \times 9.8\,\text{m s}^{-2})$
$= (R - 490\,\text{N})$

$(R - 490\,\text{N}) = 50\,\text{kg} \times 5.0\,\text{s}^{-2}$
$R = 490\,\text{N} + 250\,\text{N} = 740\,\text{N}$

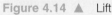

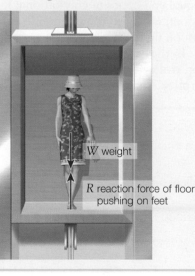

T

W

Figure 4.14 ▲ Lift

W weight

R reaction force of floor
pushing on feet

Figure 4.15 ▶
Free body force diagram of woman in lift

Systems of interacting bodies

In order to examine the motion of a single body, we isolated the forces acting on the body and ignored the effect of the forces on surrounding objects. When more than one body is considered, the interacting objects make up a **system**. On a large scale, we have the Solar System, in which the Sun and planets interact by gravitational forces; on a smaller scale, two colliding snooker balls can be considered as a system in which contact forces prevail. Most of the systems studied at AS level will involve only two bodies.

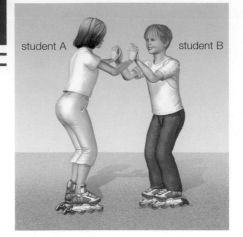

Figure 4.16 ▲
Who is pushing?

Definition
Newton's third law of motion states that if body A exerts a force on body B, body B will exert an equal and opposite force on body A.

4.4 Newton's third law of motion

If you push against a wall, you will feel the wall pushing back on your hands. Imagine you are in space close to the stationary rock described on page 30. The rock and you will remain at rest unless either is subjected to a resultant force. If you push the rock, it will move away from you, but you will also 'feel' a force pushing back on your hands. This force will change your state of rest and you will move away from the rock.

The following experiment can be used to test this principle on Earth. Two students on roller skates face each other on a level surface.

First, student A pushes student B (the rock) and then student B pushes student A. In each case, both students will move away from the other.

Sir Isaac Newton used the same 'thought experiment' to formulate his third law of motion: if body A exerts a force on body B, body B will exert an equal and opposite force on body A.

Newton's third law pairs

Newton's third law always applies to a pair of objects. Consider the Moon orbiting the Earth.

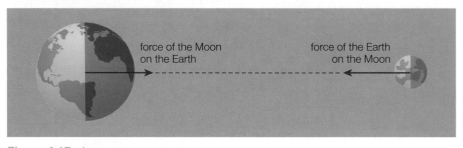

Figure 4.17 ▲
The Earth and the Moon

The Earth exerts a **gravitational** force **on the Moon**, and the Moon exerts an **equal** and **opposite gravitational** force **on the Earth**. The single force on the Moon is needed to maintain its orbit around the Earth, while the force of the Moon on the Earth gives rise to the tides.

In the case of the trolleys, trolley A exerts a **contact** (compression) force **on trolley B** and **trolley B** exerts an **equal** and **opposite contact** force **on trolley A**.

Experiment

Illustrating Newton's third law

A spring is attached to trolley A. An identical trolley, trolley B, is connected to trolley A using a fine thread. The spring is compressed as shown in Figure 4.18. The trolleys are placed on a smooth, horizontal surface.

The thread is burned and the trolleys move apart. Assuming that the surfaces are uniform and the trolleys roll in a similar manner, the distances travelled will be proportional to the applied force.

The experiment is repeated with the spring attached to trolley B. In both cases, the trolleys should move the same distance.

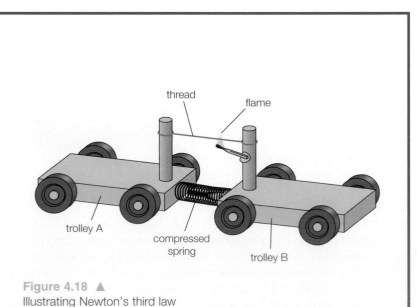

Figure 4.18 ▲
Illustrating Newton's third law

Newton's third law pairs must always:

- act on two separate bodies
- be of the same type
- act along the same line
- be equal in magnitude
- act in opposite directions.

Other examples of Newton's third law include a person standing on Earth and rocket propulsion.

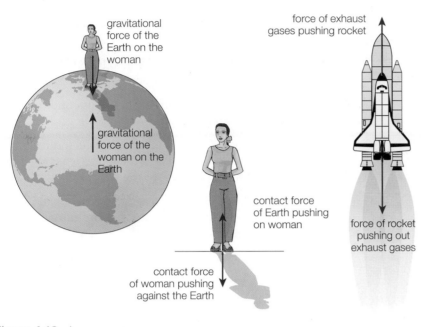

Figure 4.19 ▲
Some Newton's third law pairs

The woman and the Earth experience two pairs of forces. The Earth pulls the woman down with a gravitational force and the woman pulls the Earth up with an equal gravitational force. But there is also a pair of contact forces at the ground. The woman pushes down onto the surface of the Earth and the ground pushes up on her feet with an equal and opposite force.

To expel the exhaust gases the rocket must apply a force on the gases, pushing them backward. The gases must therefore exert an equal and opposite force on the rocket, driving it forward.

Newton's third law and equilibrium

Let us consider forces in equilibrium. It is apparent that an object at rest on the ground is subjected to a downward gravitational force (its weight) and an equal upward contact force (the normal reaction). The object is in equilibrium as the resultant force is clearly zero. The rocket, however, is subjected to a single force and is obviously not in equilibrium. In deep space, the centre of mass of the rocket **and** the fuel remain fixed, and in the absence of any external forces, the **system** (rocket plus fuel) is in equilibrium.

> **Tip**
>
> Never try to apply Newton's third law to a single body.

REVIEW QUESTIONS

1 A car of mass 1500 kg tows a trailer of mass 2500 kg. The driving force of the road on the wheels of the car is 7000 N and both the car and the trailer experience resistive forces of 1000 N.

   a) The acceleration of the vehicles is:

   A 1.25 m s^{-2} B 1.75 m s^{-2}

   C 2.00 m s^{-2} D 3.33 m s^{-2}

   b) The tension in the coupling when the driving force is reduced so that the acceleration falls to 1.00 m s^{-2} is:

   A 2500 N   B 3500 N

   C 4000 N   D 6000 N

2

Figure 4.20 ▲

   a) The weight of the man in Figure 4.20 makes a Newton's third law pair with which other force?

   A The downward contact force of the man on the table

   B The upward contact force of the table on the man

   C The upward gravitational force of the man on the Earth

   D The upward gravitational force of the man on the table

   b) The man is in equilibrium because his weight is equal to:

   A the downward contact force of the man on the table

   B the upward contact force of the table on the man

   C the upward gravitational force of the man on the Earth

   D the upward gravitational force of the man on the table.

3 Explain the differences between forces at a distance and contact forces. Give one example of each.

4 Complete Table 4.2 by including one example of each type of force.

Type of force	Example
Gravitational	
Electromagnetic	
Nuclear	

Table 4.2 ▲

5 a) What condition applies for a body to be in equilibrium under the action of several forces?

   b) Describe an experiment to show this condition for three vertical forces.

6 A boat of mass 800 kg is pulled along the sand at constant velocity by a cable attached to a winch. The cable is at an angle of 15° to the horizontal.

   a) Draw a free-body force diagram for the boat, labelling the forces **tension**, **friction**, **normal reaction** and **weight**.

   b) State which of the forces acts over a distance.

   c) If the tension in the cable is 4.0 kN, show that:

   i) the frictional force opposing the motion of the boat is about 3.9 kN

   ii) the normal reaction of the sand on the boat is about 7.0 kN.

7 a) State Newton's third law of motion.

   b) Figure 4.21 shows a satellite in orbit around the Earth.

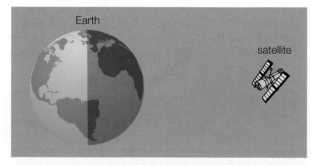

Figure 4.21 ▲

   i) Copy the diagram and show the forces acting on each body.

   ii) State three properties of these forces that show them to be a Newton's third law pair.

5 Work, energy and power

We continually use energy in everyday life. Energy is needed to move around, to keep us warm and to manufacture things. But what is it and where does it go? In this section, you will investigate and apply the principle of conservation of energy, including the use of work done, gravitational potential energy and kinetic energy.
You will also look into some mechanical applications of work, energy and power.

5.1 Work and energy

Work

Lift this book about 50 cm above the table. You have done some work. Now lift the book about one metre above the table. You have done more work. If two books are used, even more work will be done. The amount of work you do depends on the force applied to lift the books and how far it is moved. Work is done when the point of application of a force is moved in the direction of the force:

Work done = force × distance moved by the force in the direction of the force

Work is measured in joules (J).

Worked example

Calculate the work done in raising a book of mass 0.80 kg through a height of 1.50 m.

Answer

Force applied to raise the book = $0.80 \, \text{kg} \times 9.8 \, \text{m s}^{-2} = 7.8 \, \text{N}$

Work done by the force = $7.8 \, \text{N} \times 1.50 \, \text{m} = 12 \, \text{J}$

A force often moves an object in a different direction to the direction of the applied force.

Figure 5.1 ▲

The work done by the child pulling the pram in Figure 5.1 is the product of the horizontal component of the force and the horizontal displacement, Δx, of the pram:

Work done on pram = $F \cos \theta \times \Delta x$

This may also be expressed as the product of the force and the component of the displacement along the line of action of the force:

Work done on the pram = $F \times \Delta x \cos \theta$

In this example, the angle θ must be between $0°$ and $90°$. The value of $\cos\theta$ will be between $+1$ and 0, and hence the work done will be positive. A value of θ between $90°$ and $180°$ would indicate that the force is acting in the opposite direction to the movement of the pram. Work is done **by** the pram against the resistive forces, so $F\cos\theta$ will be negative.

In most cases, the forces acting on moving objects are not constant, and it is usual to express the work done in terms of the average force:

$$\Delta W = F_{ave}\,\Delta x$$

Force and displacement are **vector quantities**, and you might expect that their product should also be a vector. However, if a force is moved first in one direction and then by the same distance in the opposite direction, the vector sum of the work done would equal zero. It therefore follows that work done has size but no direction and is therefore a **scalar quantity**.

Relationship between work and energy

Work and energy are closely linked. Energy is transferred **to** an object when work is done on it. When energy is transferred **from** the object to another system we say that the object does work.

When you raised the book off the bench, you transferred energy to the book. To stop a moving bicycle by applying the brakes, energy is transferred from the bike to the braking system.

We have already defined work, so it is convenient to define energy in terms of work done: an object has energy when it has the ability to do work. This statement is useful, but it is not universal. Heating is energy transfer between regions of different temperatures. In heat engines, some but not all of the energy transferred can do work. Heat energy will be studied in more detail in the A2 course. It follows that energy, like work, is measured in joules (J).

All forms of energy can be described in terms of **potential energy** and **kinetic energy**.

Potential energy

In Figure 5.2a, work is done on the box when it is lifted onto the bench. The upward force needed to raise the box has the same magnitude as its weight (mg).

Work done to lift the box = force × distance moved

$$\Delta W = mg \times \Delta h$$

> **Definition**
>
> An object has energy when it has the ability to do work

> **Definition**
>
> Potential energy is the ability of an object to do work by virtue of its position or state.

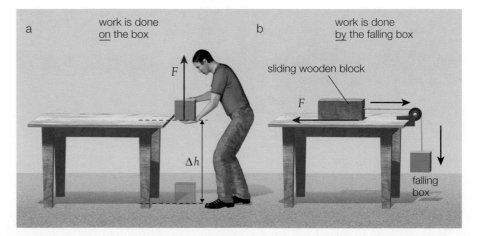

a work is done on the box

b work is done by the falling box

sliding wooden block

F

Δh

F

falling box

Figure 5.2 ▶

Imagine the box is attached to a block of wood by a length of string, as shown in Figure 5.2b. If the box falls off the bench, the pull of the string on the block will do work as the block moves along the bench. The box on the bench has the **potential** to do work by virtue of its position. Because the work is done by the displacement of a gravitational force, the box is said to have **gravitational potential energy (GPE).**

Whenever the box is raised by a height Δh, its *GPE* is increased. The gain in *GPE* (ΔGPE) is equal to the work done on the box:

$$\Delta GPE = m\, g\, \Delta h$$

If the box is raised more than a few kilometres above the surface of the Earth, the value of the gravitational field strength, g, will become noticeably smaller. The value will continue to fall as the distance from the Earth increases. The expression above therefore relates to variations in height close to the Earth's surface, where g can be assumed to be constant ($9.8\,\mathrm{N\,kg^{-1}}$).

In Figure 5.3, work is done stretching the rubber band. When the band is released, the weight slides across the surface doing work against the frictional force. Energy has been stored in the rubber band, which enables it to do work. This is called **elastic potential energy (EPE).**

Definition

Elastic potential energy (strain energy) is the ability of an object to do work by virtue of a change in its shape.

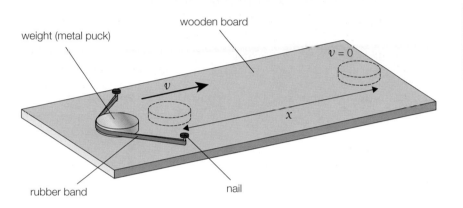

Figure 5.3 ▲

The *EPE* of a deformed object can be represented in terms of the work done to change its shape. For a spring or wire being stretched by an extension Δx

$$\Delta EPE = F_{ave}\, \Delta x$$

Elastic potential energy will be further investigated in Chapter 7.

Kinetic energy

Anything that moves is able to do work. Imagine a hammer driving a nail into a wall. Just before impact, the hammerhead is moving quickly. When the nail is struck, it moves into the wall and the hammer stops. The hammer has transferred energy to do work driving the nail into the wall.

Definition

Kinetic energy is the ability of an object to do work by virtue of its motion.

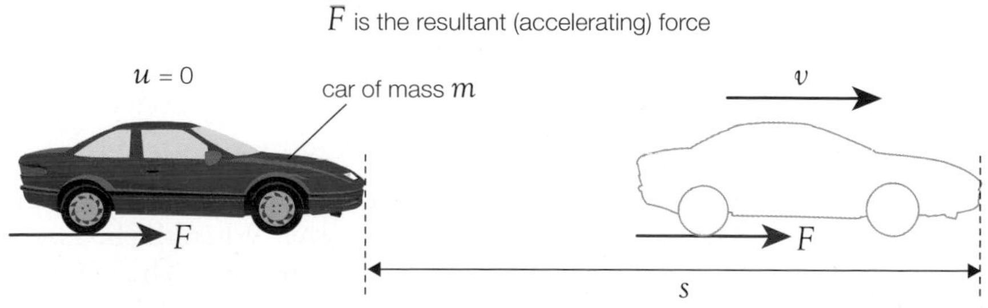

An expression for the kinetic energy (*KE*) of a moving object can be found using Newton's laws (Sections 4.2–4.4) and the equations of uniformly accelerated motion (Section 3.1).

In Figure 5.4, a constant force, F, is applied to the car over distance s. The car accelerates from rest ($u = 0$) until it reaches velocity v.

F is the resultant (accelerating) force

$u = 0$
car of mass m

v

F

F

s

Figure 5.4 ▲

Gain in KE = work done on the car = $F \times s = ma \times s = m \times as$

Using the equation:

$$v^2 = u^2 + 2as = 0 + 2as$$

$$\Rightarrow as = \frac{v^2}{2}$$

Gain in $KE = m \times \dfrac{v^2}{2} = \frac{1}{2}mv^2$

The kinetic energy of an object of mass m moving with a speed v is $\frac{1}{2}mv^2$

Worked example

Estimate the kinetic energy of an Olympic sprinter crossing the finish line.

Answer
When an estimate is required, a wide range of answers will be accepted. You will be required to make sensible approximations of the mass and speed of the athlete.

Mass of male athlete = 80 kg (180 lb)
Mass of female athlete = 60 kg (130 lb)

Time for 100 m is around 10 s (11 s for women), so a reasonable estimate of speed = $10\,\text{m s}^{-1}$.

Kinetic energy = $\frac{1}{2}mv^2$

$= \frac{1}{2} \times 60\,\text{kg} \times (10\,\text{m s}^{-1})^2$

$= 3000\,\text{J}$

$= 3\,\text{kJ}$

> **Tip**
>
> As such a wide range of values is possible with estimates of this type, estimates should be written to one significant figure. If you had chosen $m = 65\,\text{kg}$ and $v = 11\,\text{m s}^{-1}$, the kinetic energy would be 3932.5 J but should be given as 4000 J or 4 kJ.

If a mass changes speed from u to v:

change in kinetic energy, $\Delta KE = \frac{1}{2}mv^2 - \frac{1}{2}mu^2$

Worked example

A car of mass 1500 kg travelling at $20\,\text{m s}^{-1}$ is slowed down to $10\,\text{m s}^{-1}$ by applying the brakes. The car travels 30 m during braking.

1. Determine the kinetic energy transferred from the car.

2. Calculate the average resistive force acting on the car.

Answer

1. KE transferred = $\frac{1}{2} \times 1500\,\text{kg} \times (20\,\text{m s}^{-1})^2 - \frac{1}{2} \times 1500\,\text{kg} \times (10\,\text{m s}^{-1})^2$
   $= 2.25 \times 10^5\,\text{J}$
   $= 2.3 \times 10^5\,\text{J}$

2. Work done = force × distance
   $= F \times 30\,\text{m}$
   $= 2.25 \times 10^5\,\text{J}$
   Average force = $7.5 \times 10^3\,\text{N}$

> **Tip**
>
> When all the quantities are given to two significant figures, the final answer should also be given to two significant figures. However, intermediate calculations should not be rounded: e.g. in the Worked example $2.25 \times 10^5\,\text{J}$ should be used in the calculation of average force.

Gravitational potential energy

When the pendulum bob in Figure 5.5 is displaced to one extreme, it is raised by a height Δh. The *GPE* of the bob increases by $mg\Delta h$. When the bob is released, it begins to fall and accelerates towards the midpoint. The moving bob gains kinetic energy. The *GPE* of the bob decreases to a minimum value at the lowest point of the swing where the *KE* has a maximum value. The *KE* is then transferred back to *GPE* as the bob moves up to the other extreme.

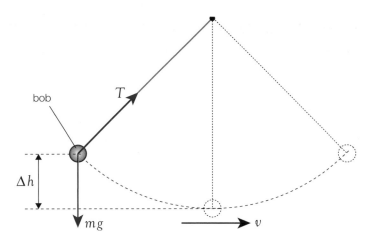

 Figure 5.5 ◀

The motion of the pendulum can be described as a continuous variation of GPE and KE:

$$\Delta GPE \rightarrow \Delta KE \rightarrow \Delta GPE$$
$$mg\,\Delta h \rightarrow \tfrac{1}{2}mv^2 \rightarrow mg\,\Delta h$$

All oscillators show a similar variation between *PE* and *KE*. For masses on springs or rubber bands, elastic potential energy stored at the ends of the oscillation is transferred to kinetic energy at the midpoint.

Internal energy

In many transfers, energy seems to be lost to the surroundings. A pendulum will swing with decreasing amplitude as work is done against air resistance and friction at the support. The brakes of a bicycle get hot when *KE* is transferred to them. Both of these conversions lead to an increase in the kinetic and potential energies of the particles (atoms or molecules) in the air and the brakes. This is often, wrongly, described as heat energy.

In a gas of fixed volume, the internal energy is the sum of the *KE* of all the molecules. In a solid, the atoms vibrate with continuously varying *PE* and *KE*.

An increase in internal energy usually results in a rise in temperature. Thermal energy and internal energy are studied in greater detail in the A2 course.

Experiment

Investigating the transfer of *GPE* to *KE*

A trolley is held at the top of a track, as shown in Figure 5.6. The difference in the height, Δh, of the centre of gravity of the trolley (its midpoint) when it is at the top of the ramp and when it is on the horizontal board is measured.

The trolley is allowed to run down the ramp onto the horizontal board, and it cuts through the light gate. The speed, v, of the trolley and masses is found by measuring the time, Δt, for the card (width Δx) to cross the light gate:

$$v = \frac{\Delta x}{\Delta t}$$

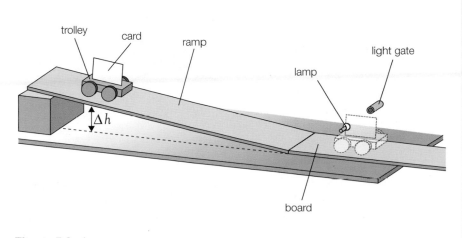

Figure 5.6 ▲

The experiment is repeated several times and an average value of v is found. If the mass of the trolley is known, the loss in potential energy and the gain in kinetic energy can be calculated.

ΔGPE of the trolley $= m g \, \Delta h$

ΔKE of the trolley $= \frac{1}{2} m v^2$

Theoretically, the values of ΔGPE and ΔKE should be equal, but experimental uncertainties and work done against air resistance and the friction in the pulley are likely to affect the readings.

The energy transfers can be further investigated using a range of values of Δh and measuring the average velocities. A graph of ΔKE against ΔGPE is then drawn. The percentage of the original GPE converted to KE is found from the gradient of the graph.

A set of results taken from the experiment is as follows:

Mass of trolley, $M = 0.400\,$kg

Length of card, $x = 0.200\,$m

1 Complete Table 5.1 (use an Excel or Lotus 123 spreadsheet if available).

a) Plot a graph of ΔKE/J against ΔGPE/J.

b) Is there a systematic difference between ΔKE and ΔGPE?

c) Find the gradient of the graph; this will give you the fraction of the GPE that is converted into KE. Why do you expect the value to be less than one?

Δh/m	ΔGPE/J	t/s	v/m s⁻¹	ΔKE/J
0.05		0.22		
0.10		0.16		
0.15		0.13		
0.20		0.11		
0.25		0.10		

Table 5.1 ▲

An alternative method to observe the conversion of GPE to KE is to allow an object to fall from a range of heights and measure the time taken for the object to pass through a light beam. The ΔGPE is then compared with ΔKE as for the previous experimental set-up.

(The uncertainties of the readings will be improved if several values of time are taken for each height. See Section 2.2 for more details of uncertainties.)

Worked example

The car on the funfair ride in Figure 5.7 has mass m and is released from rest at the top of the track, which is at height h above the lowest level of the ride. The car reaches speed v at the bottom of the dip.

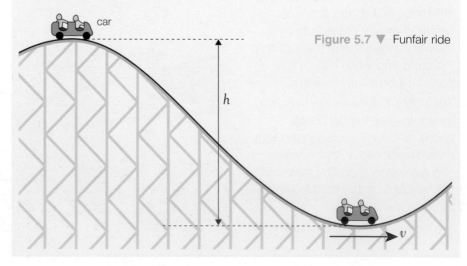

Figure 5.7 ▼ Funfair ride

1 Write down expressions for:
  a) the gravitational potential energy transferred from the car as it runs from top to bottom
  b) the kinetic energy of the car at the lowest point.

2 If h is 20 m, show that the speed of the car should be about 20 m s^{-1}.

3 The speed of the car is measured as 19 m s^{-1}. Explain why this value is lower than the value calculated in Question 2.

4 Do you expect that the car will pass the next peak, which is at a height of 15 m above the dip? Explain your reasoning.

Answer

1 a) $GPE = mgh$
  b) $KE = \frac{1}{2}mv^2$

2 $mgh = \frac{1}{2}mv^2 \Rightarrow v = \sqrt{(2gh)}$
$$= \sqrt{(2 \times [9.81 \text{ m s}^{-2}] [20 \text{ m}])}$$
$$= 19.8 \text{ m s}^{-1} \approx 20 \text{ m s}^{-1}$$

3 The car does work against frictional forces and air resistance. Some of the GPE is transferred to increase the internal energy of the surroundings.

4 The car needs to transfer 75% of its original GPE to reach a height of 15 m. The energy transferred from the car to the surroundings on the downward journey is less than 10% of the GPE, so it is likely that the car will reach the top of the hill.

This section is mainly concerned with **mechanical energy**. The potential and kinetic energy of objects have been explained in terms of mechanical work. Other forms of energy include:

- **Chemical energy** – the ability to do work using chemical reactions (fuels and electric cells store chemical energy).
- **Nuclear energy** – the ability to do work by changes in the constitution of nuclei (nuclear fission and fusion are examples where energy stored in the nucleus can be transferred to other forms).
- **Electrostatic potential energy** – the energy of a charged particle by virtue of its position in an electric field (charged capacitors store electrostatic energy).
- **Radiant energy** – energy transferred by electromagnetic waves (radio waves, light and X-rays are examples of radiant energy).

5.2 Principle of conservation of energy

We have seen that energy is readily transferred from one form to another by work. Energy can also be transferred by heating. Often it seems that energy has been lost from a system. The examples of the pendulum and bicycle on page 43 illustrate that although some forms of energy have gone from the systems, they have simply been transferred to the surroundings in a different form.

Energy is never created or destroyed, but it can be transferred from one form into another. This principle is universal. The Sun and other stars transfer nuclear energy to radiant energy in the form of electromagnetic waves. The Earth receives some of this energy (noticeably as light and infrared radiation) and also radiates energy back into space.

In nuclear reactions, such as alpha and beta decay, some mass is transferred to the particles in the form of kinetic energy.

> **Tip**
>
> Energy cannot be created or destroyed, it can only be transferred from one form to another.

A thermal power station illustrates the conservation of energy. The aim is to convert the chemical energy in the fuel (gas, oil or coal) into electrical energy. A simplified version of the transfers is:

chemical energy in the fuel → internal energy of the compressed steam → electrical energy

A large amount of internal energy is transferred from the low pressure steam to the cooling water in the turbine condenser. This is usually dissipated to the atmosphere, which results in an increase in the internal energy of the environment.

Energy transfers can also be represented in Sankey diagrams like that in Figure 5.8. The width of the arrow at the input represents the chemical energy in the fuel. The widths of the outgoing arrows represent the energy transfers. Note that the sum of the widths of the outputs equals that of the input. This indicates that the energy is conserved.

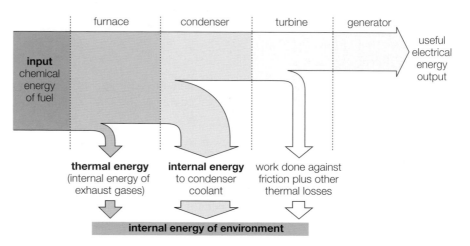

Figure 5.8 ▶

Efficiency

The Sankey diagram shows that only a fraction of the chemical energy is converted into electrical energy. The ability to transfer the input energy to desired (useful) energy is the efficiency of the system.

$$\text{Efficiency} = \frac{\text{useful energy output}}{\text{energy input}} \times 100\%$$

Worked example

A pulley system is used to lift a 100 kg mass to a height of 1.5 m. The operator applies a force of 200 N and pulls the rope a total distance of 8.0 m.

Calculate:

a) the gain in GPE of the load (useful energy output)

b) the work done by the operator (energy input)

c) the efficiency of the system.

Answer

a) $\Delta GPE = mg\Delta h$
$$= 100\,\text{kg} \times 9.8\ \text{m s}^{-2} \times 1.5\,\text{m}$$
$$= 1470\,\text{J}$$

b) $\Delta W = F \times x$
$$= 200\,\text{N} \times 8.0\,\text{m}$$
$$= 1600\,\text{J}$$

c) $\text{Efficiency} = \dfrac{1470\,\text{J}}{1600\,\text{J}} \times 100\% = 92\%$

5.3 Power

A car engine transfers the chemical energy from the fuel to the work needed to move the car. If two cars use the same amount of fuel, an equal amount of chemical energy will be transferred to work by the engines. The more powerful car will use the fuel more quickly.

An electric lamp converts electrical energy into radiant (light) energy. A low power lamp is much dimmer than a high power one, but if it is left on for a longer time it can emit more energy. Both examples indicate that power depends on the rate at which energy is transferred:

$$\text{Power} = \frac{\text{work done (energy transferred)}}{\text{time taken}}$$

Power is measured in **watts (W)**. One watt is a rate of conversion of one joule per second:

$$1\,\text{W} = 1\,\text{J}\,\text{s}^{-1}$$

Definition

Power is the rate of doing work or of energy conversion.

Worked example

Estimate the minimum power you would need to generate to climb a hill of height 80 m in 5 minutes.

Answer

$$\begin{aligned}
\Delta W &= \Delta GPE \\
&= mg\,\Delta h \\
&= (\text{your mass kg}) \times 9.8\,\text{m}\,\text{s}^{-2} \times 80\,\text{m}
\end{aligned}$$

For $m = 70\,\text{kg}$:

$$\begin{aligned}
\Delta W &= 70\,\text{kg} \times 9.8\,\text{m}\,\text{s}^{-2} \times 80\,\text{m} \\
&\approx 56\,000\,\text{J}
\end{aligned}$$

$$\text{Power} = \frac{\Delta W}{\Delta t} \approx \frac{56\,000\,\text{J}}{5 \times 60\,\text{s}} \approx 200\,\text{W}$$

Experiment

Measuring the output power of an electric motor

The motor in Figure 5.9 is switched on until the load is raised almost up to the pulley wheel and is then switched off. A stopclock is used to measure the time, Δt, when the motor is working, and a metre rule is used to find the distance, Δh, through which the load is lifted. The experiment is repeated for a range of masses to investigate the effect of the load on the power output of the motor using the equation:

$$\text{Power output of motor} = \frac{\Delta GPE}{\Delta t} = \frac{mg\,\Delta h}{\Delta t}$$

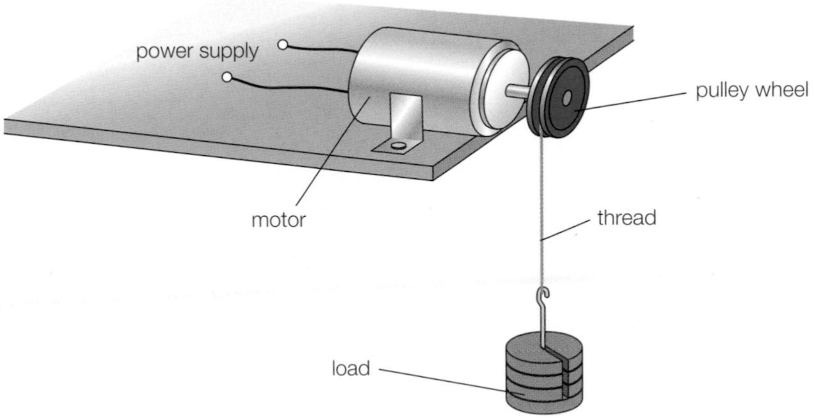

Figure 5.9 ▲

In Chapter 11, power in electrical circuits will be introduced. A similar experiment will be performed to measure the input power to the motor and hence the efficiency of the motor.

Power and motion

If a canoeist paddles at speed v for distance Δx against a uniform retarding force F, she will be working at a constant rate:

$$\Delta W = F \times \Delta x$$

$$\text{Power} = \frac{\Delta W}{\Delta t} = \frac{(F \times \Delta x)}{\Delta t} = F \times \frac{\Delta x}{\Delta t} = F \times v$$

Power (W) = retarding force (N) × speed (m s^{-1})

$$P = Fv$$

Worked example

A motorised wheelchair is driven at 2.5 m s^{-1} against an average resistive force of 80 N. Calculate the efficiency of the wheelchair if its motor has a power rating of 250 W.

Answer

Useful power output, $P_o = 80\,\text{N} \times 2.5\,\text{m s}^{-1} = 200\,\text{W}$

$$\text{Efficiency} = \frac{\text{useful power output}}{\text{power input}} \times 100\%$$

$$= \frac{200\,\text{W}}{250\,\text{W}} \times 100\% = 80\%$$

REVIEW QUESTIONS

1 A barge is pulled along a canal by two men who each apply a force of 100 N to a rope at 60° to the banks.

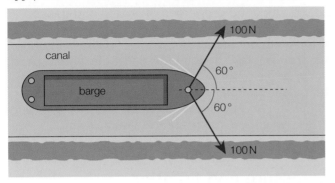

Figure 5.10 ▲

The total work done by the men in moving the barge 100 m along the canal is:

A 5000 J B 8660 J

C 10 000 J D 17 400 J

2 A woman performs a 'bungee jump', falling from a bridge with a length of elastic cord tied around her ankles. The gravitational potential energy lost by the jumper at the lowest point has been transferred as:

A elastic strain energy in the cord

B elastic strain energy and internal energy in the cord

C kinetic energy of the jumper

D kinetic energy of the jumper plus internal energy in the cord.

3 A forklift truck raises ten 100 kg sacks of rice off the floor onto a storage shelf 2.4 m high in 8.0 s. The average power generated by the truck is about:

A 300 W B 2400 W

C 3000 W D 24 000 W

4 A toy tractor with a 120 W electric motor is driven at a constant speed of 2.0 m s^{-1}. If the motor is 75% efficient, the average resistive force on the tractor is:

A 45 N B 60 N

C 180 N D 240 N

5 Define work and state the unit of work.

6 a) A husky dog pulls a sleigh along a level stretch of snow over a distance of 100 m. If the average force exerted on the sleigh is 60 N, how much work is done by the dog?

b) The sleigh now moves across a slope so that the husky pulls with an average force of 80 N at an angle of 30° up the slope from the direction of travel of the sleigh. How much work does the dog do to move the sleigh 50 m along this path?

7 a) What is potential energy?

b) Explain the difference between gravitational and elastic potential energy.

8 a) What is kinetic energy?

b) Calculate the kinetic energy of a golf ball of mass 50 g travelling at 20 m s^{-1}.

9 A ramp is set up in the laboratory so that a trolley is able to run from the top to the bottom. Describe an experiment you could perform to find out what percentage of the gravitational potential energy lost is converted to kinetic energy. Your answer should include:
● a diagram including any additional apparatus required
● a description of how the apparatus is used
● a description of how the results are analysed.

10 State the meaning of the efficiency of a system.

11 a) Define power and state the unit of power.

b) A student of mass 70 kg runs up a flight of stairs in 5.0 s. If there are 20 steps each of height 25 cm, calculate the average power needed to climb the stairs in this time.

c) Explain why the power generated by the student is likely to be greater than this value.

12 A cyclist pedals along a level road at 5.0 m s^{-1}. The rider stops pedalling and 'free wheels' until he comes to rest 20 m further down the road.

a) Show that the deceleration of the bike is approximately 0.6 m s^{-2}.

b) If the mass of the bike plus rider is 100 kg, calculate:

i) the average frictional force opposing the motion

ii) the power needed for the cyclist to maintain a steady speed of 5.0 m s^{-1}.

6 Fluids

A fluid is a material that flows. Unlike a solid, in which the atoms occupy fixed positions, the particles of a fluid can move relative to one another.
In this section, you will study the properties of liquids and gases, which are naturally fluid. You will learn about substances like chocolate and margarine to understand how their fluidity is important in the food industry. You will also become aware of the significance of moving fluids in car and aircraft design and the transportation of oil and gas through pipelines. Finally, you will learn about blood, which is a fluid vital to life, and the factors that affect its flow through our veins and arteries, which are important for healthy living.

6.1 Density, pressure and flotation

Density of fluids

Liquids and gases expand much more than solids when they are heated, so a fixed mass of fluid occupies a bigger volume than the solid form and so its density is reduced. Liquids are generally considered to be incompressible, but gases are readily squeezed (try putting your finger over the outlet of a bicycle pump and pushing in the handle). Because of this, the pressure needs to be stated in addition to the temperature when the density of a gas is quoted.

Table 6.1 gives some examples of the densities of fluids. The values are at $293\,\text{K}$ and gas pressure of $1.01 \times 10^5\,\text{Pa}$.

Fluid	Density/kg m^{-3}
Mercury	13 600
Water	1000
Ethanol	790
Carbon dioxide	1.78
Air	1.24
Helium	0.161
Hydrogen	0.081

Table 6.1 ▲
Densities of some fluids

Definition

$$\text{Density} = \frac{\text{mass}}{\text{volume}}$$

$$\rho = \frac{m}{V}$$

Experiment

Finding the density of air

A flask and its attachments are placed onto a balance (sensitivity ±0.01 g or less) and the total mass is recorded. A vacuum pump is used to remove as much air as possible from the flask (Figure 6.1). The flask should be encased with a stiff wire mesh as a precaution against implosion and safety goggles must be worn. A protective screen between the flask and observers is also recommended. The flask and attachments are reweighed so that the mass of the gas removed from the flask can be found.

To measure the volume of gas (at its initial pressure), the end of the rubber tube is immersed in a beaker of water and the clip is released. Water is forced into the tube by the external air pressure, and the volume of the water in the flask equals the volume of the evacuated air.

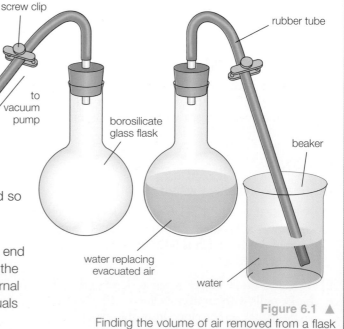

Figure 6.1 ▲
Finding the volume of air removed from a flask

Pressure in fluids

If you dive to the bottom of a swimming pool, you will feel the water pressure pushing into your ears. This pressure is created by the weight of the water above you, and the deeper you go, the greater this pressure. Similarly, the weight of the atmosphere produces an air pressure of about 1.0×10^5 Pa at the Earth's surface.

Consider a column of fluid of height h, density ρ and area of cross-section A.

$$\text{Pressure at the base} = \frac{\text{weight of column}}{\text{area}} = \frac{mg}{A} = \frac{V\rho g}{A}$$

Volume, V, of the column $= Ah$

Hence, $P = h\rho g$

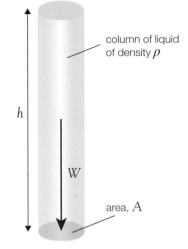

column of liquid of density ρ

Figure 6.2 ▲

For large values of h, gases compress in the lower regions. This means that the Earth's atmosphere has a lower density at higher altitudes, and its pressure is therefore not directly proportional to height above Earth.

Worked example

Estimate the height of the Earth's atmosphere.

Answer
The density of air is about 1.24 kg m^{-3} at sea level and virtually zero at upper levels of the atmosphere, so the average density can be taken as 0.62 kg m^{-3}. If air pressure is assumed to be 1.0×10^5 Pa and g to be 10 N kg^{-1}:

$$h = \frac{P}{\rho g} = \frac{1.0 \times 10^5 \text{ Pa}}{0.62 \text{ kg m}^{-3} \times 10 \text{ N kg}^{-1}} \approx 16 \text{ km}$$

Upthrust in fluids

If you are in a swimming pool, you will experience a buoyancy force that enables you to float or swim. This force is called an **upthrust**, and it is a consequence of the water pressure being greater below an immersed object than above it.

The upthrust is the difference between the force due to water pressure at the bottom of the cylinder, F_1, and that at the top, F_2.

For a fluid of density ρ:

$F_1 = P_1 A = h_1 \rho g A$

$F_2 = P_2 A = h_2 \rho g A$

$U = F_2 - F_1 = (h_2 - h_1)\rho g A = (h_2 - h_1)A\rho g = V\rho g = mg$

The upthrust is equal to the **weight** of the displaced fluid.

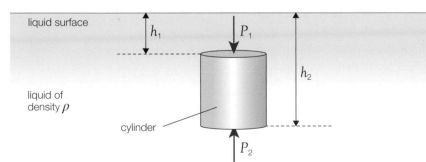

Figure 6.3 ◄

This result is often stated as **Archimedes' principle**.

Archimedes (circa 287–212 BC) was seeking a method to verify that the king's crown was made from pure gold and is reported to have leaped from his bath and run through the streets shouting 'Eureka' (I've found it!).

Estimate the upthrust on your body when it is totally immersed in water and compare it with the upthrust experienced due to the air you displace. You can consider your head as a sphere and your trunk, arms and legs as cylinders.

Flotation

An object will float in a fluid if the upthrust – that is, the weight of the fluid it displaces – is equal to its weight.

Figure 6.4 shows a fully laden ship floating in the cold salt water of the North Atlantic Ocean and the same ship in the warmer fresh water of an upriver tropical port. The ship needs to displace more of the less dense freshwater to balance its weight, and so lies deeper in the fresh water than in the salt water.

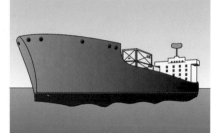

vessel in cold salt water   vessel in warm fresh water

Figure 6.4 ▲

Insurance companies require all cargo ships to have maximum load levels on their hulls, and a number of such lines are painted on vessels used for international trade to represent the safety levels in different waters. These lines were introduced as law in the UK in the late nineteenth century by the Member of Parliament Samuel Plimsoll, and are still referred to as 'Plimsoll lines'.

Worked example

Calculate the density of the hot air in a balloon floating at a fixed height close to the ground. The density of the cold air is $1.4 \, \text{kg m}^{-3}$. The total mass of the balloon's fabric, gondola, fuel, burners and occupants is 700 kg, and its volume is $2500 \, \text{m}^3$.

Answer

Upthrust, U = weight of displaced air
$$= 2500 \, \text{m}^3 \times 1.4 \, \text{kg m}^{-3} \times 9.8 \, \text{N kg}^{-1}$$
$$= 34\,000 \, \text{N}$$

For the balloon to be in equilibrium:

U = weight of balloon, occupants and accessories + weight of hot air

Weight of hot air = $35\,000 \, \text{N} - 700 \, \text{kg} \times 9.8 \, \text{N kg}^{-1} = 27\,000 \, \text{N}$

Mass of hot air = 2700 kg

Density of hot air = $\dfrac{2700 \, \text{kg}}{2500 \, \text{m}^3} = 1.1 \, \text{kg m}^{-3}$

Archimedes' ideas are still in common use in industry, where the principle of flotation is applied to determine the densities of solids and fluids.

6.2 Moving fluids – streamlines and laminar flow

Streamlines represent the velocity of a fluid at each point within it. They can be drawn as arrowed lines that show the paths taken by small regions of the fluid.

Figure 6.5a represents the flow of air relative to an aircraft wing and Figure 6.5b the flow of ink from a nozzle into water flowing through a pipe. In Figure 6.5a, the air above and below the aircraft wing exhibits **laminar flow** – that is, adjacent layers of air do not cross into each other. Beyond the wing, the air swirls around and forms **vortices** or **eddy currents**. The streamlines are no longer continuous and the flow is said to be **turbulent**. Figure 6.5b shows that the flow of water in the pipe is laminar at low rates of flow, but that turbulence occurs when the rate reaches a critical level (see Figure 6.5c).

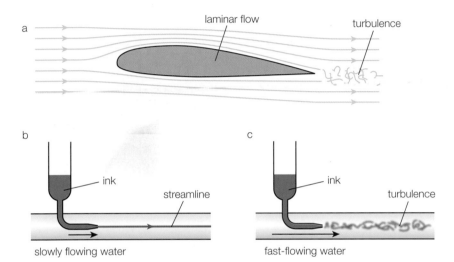

Figure 6.5 ▲

Laminar flow is an important consideration in fluid motion. The uplift on an aeroplane's wings is dependent on laminar flow, and passengers experience a rocky ride when turbulent conditions are encountered. Similarly, the drag forces on a motor car are affected by turbulence, and wind tunnels are used to observe the nature of the airflow over prototype designs.

The efficiency of fluid transfer through tubes is greatly reduced if turbulence occurs, so the rate of flow of oil and gas must be controlled so that the critical speed is not exceeded.

In the food industry, the flow of sweet casings such as toffee and chocolate over nuts or other fillings should be laminar so that air bubbles are not trapped as a result of turbulence.

Viscosity

The viscosity of a fluid relates to its stickiness and thus to its resistance to flow. Syrup and engine oil are very viscous, while runny liquids such as water and petrol and all gases have low viscosities.

Viscosity can be described in terms of the resistance between adjacent layers in laminar flow. Imagine two packs of playing cards: one brand new and the other heavily used. The cards in the new pack will slide easily over each other when pushed down at an angle from above (Figure 6.6), but the cards of the old pack will stick together, dragging the lower cards with them.

Comparing viscosities

The viscosities of liquids can be compared by observing their rates of flow through a glass tube. A simple device called a Redwood viscometer can be easily adapted for the laboratory.

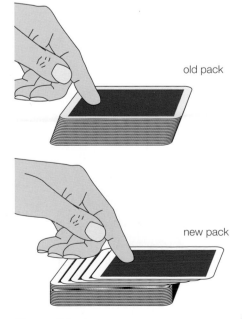

Figure 6.6 ▲

53

Comparing viscosities using a viscometer

Using the apparatus in Figure 6.7, fill the funnel with liquid to a level just above the upper mark. Open the clip to allow the liquid to flow through the tube into the beaker. Start timing when the level passes the upper mark and stop as it passes the lower one. Repeat for various samples, and then list your samples in order of increasing viscosity.

For runny liquids like water or sugar solutions of a low concentration, a capillary tube with a diameter of 1 mm should be used so that very short times (and large percentage uncertainties) are avoided. For syrup, honey, etc, tubes with bores of 5 mm or wider can be used.

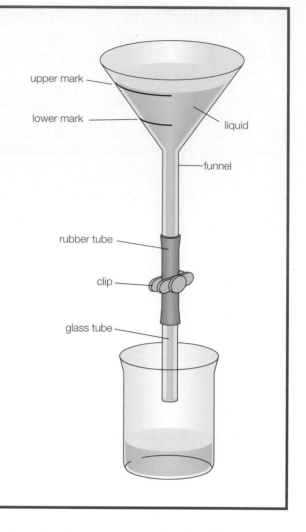

Figure 6.7 ▶

The flow of blood in our arteries and veins is important for a healthy cardiovascular system. In the nineteenth century, Jean Louis Marie Poiseuille, a French physician, showed that the rate of flow of a liquid in a uniform tube depended on the pressure per unit length across the tube, the viscosity of the liquid and the fourth power of the radius of the tube.

Although our blood vessels are not uniform pipes and the viscosity of the blood is not constant, it is apparent that the build up of fatty deposits on arterial walls from excess amounts of some forms of cholesterol will reduce the blood flow and/or increase the blood pressure. If the radius is halved, the rate of flow for a given pressure will reduce 16 times!

In many cases, patients with high blood pressure are prescribed 'blood thinners', such as aspirin, warfarin and pentoxifylline, which reduce the viscosity of the blood.

Stokes' law

When a sphere moves slowly through a fluid, the movement of the fluid relative to the sphere is laminar.

The molecules of the fluid adhere to the surface of the sphere and move along with it, creating a viscous drag between the other layers of the fluid. The Irish physicist George Gabriel Stokes deduced an expression for this force in terms of the radius of the sphere, r, the velocity relative to the fluid, v, and the **coefficient of viscosity**, η.

$$F = 6\pi\eta r v$$

This equation is generally referred to as **Stokes' law**.

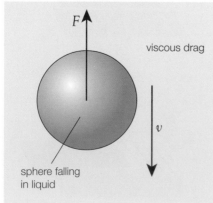

Figure 6.8 ▲

Terminal velocity

When a sphere is released and allowed to fall freely in a fluid, it is subjected to three forces: its weight, W, the upthrust, U, and the viscous drag, F.

Initially, a resultant force, $F_R = W - (U + F)$ will make the sphere accelerate downward. As the velocity of the sphere increases, the viscous drag increases according to Stokes' law until $(U + F) = W$. The resultant force then becomes zero, and the sphere continues to fall at a constant velocity known as the **terminal velocity**.

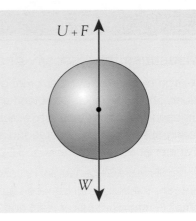

Figure 6.9 ▲
Free-body force diagram for sphere falling in a fluid

Worked example

A steel ball-bearing of mass 3.3×10^{-5} kg and radius 1.0 mm displaces 4.1×10^{-5} N of water when fully immersed. The ball is allowed to fall through water until it reaches its terminal velocity. Calculate the terminal velocity if the viscosity of the water is 1.1×10^{-3} N s m^{-2}.

Answer

$$F = W - U$$
$$6\pi\eta r v = W - U$$
$$v = \frac{W - U}{6\pi\eta r}$$
$$v = \frac{3.3 \times 10^{-5}\,\text{kg} \times 9.8\,\text{N kg}^{-1} - 4.1 \times 10^{-5}\,\text{N}}{6\pi \times 1.1 \times 10^{-3}\,\text{N s m}^{-2} \times 1.0 \times 10^{-3}\,\text{m}} = 14\,\text{m s}^{-1}$$

Measuring viscosity using Stokes' law

By measuring the terminal velocity of a sphere falling through a fluid it is possible to determine the coefficient of viscosity of the fluid.

For a sphere of radius r and density ρ_s falling through a fluid of density ρ_f and viscosity η with a terminal velocity v, the equilibrium equation:

$$U + F = W$$

can be written:

$$\tfrac{4}{3}\pi r^3 \rho_f g + 6\pi\eta r v = \tfrac{4}{3}\pi r^3 \rho_s g$$

which gives the viscosity as:

$$\eta = \frac{2(\rho_s - \rho_f)gr^2}{9v}$$

This also tells us that the terminal velocity of a falling sphere in a fluid depends on the square of its radius, so very small drops of rain – and the minute droplets from an aerosol – fall slowly through the air.

The relative viscosities of opaque liquids like molten chocolate can be compared by using a sphere with a thin rod attached so that the time interval between a mark on the lower part of the rod reaching the surface and that of a mark higher up the rod is recorded.

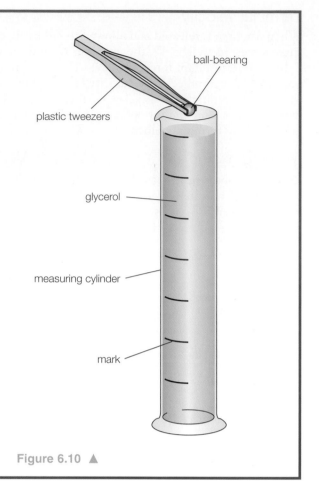

Experiment

Measuring the viscosity of a liquid

A large measuring cylinder or long glass tube is filled with a clear, sticky liquid. Glycerol works well, but syrup or honey can be used as long as they are sufficiently transparent to allow the falling spheres to be observed. Lines are drawn around the cylinder at regular intervals of 5.0 cm.

A small steel ball-bearing is released into the liquid, and the time for the ball to fall between each level is taken. If the ball is small (3 mm diameter or less) and the liquid fairly sticky, the measurements will all be similar, which indicates that terminal velocity is reached shortly after the sphere enters the liquid.

The timings are repeated and the terminal velocity is calculated by dividing the distance between the markers by the average time taken for the ball to fall between them.

The radius of the ball is found by measuring its diameter using a micrometer. Its density can be determined using the mass of one ball calculated by placing 10–100 balls on a balance. The density of the liquid also needs to be found by taking the mass of a measured volume. The viscosity is found by substituting these values in the expression:

$$\eta = \frac{2\,(\rho_s - \rho_f)\,g r^2}{9v}$$

Figure 6.10 ▲

6.3 Variations in viscosity

For most of our calculations we have assumed that the flow is laminar and that Stokes' law and Poisseiulle's equation apply. In practice, the behaviour of most fluids is much more complex. The viscosity of molten chocolate does not depend only on its temperature – it reduces as the rate of flow increases. This is an example of **thixotropy**. Many gels and colloids have this property: for example, toothpaste flows out of the tube when pressure is applied but keeps its shape on the toothbrush. Butter, margarine and other spreads normally are solid but will flow under the pressure of a knife onto the bread.

Stirring or vibration can also affect thixotropic substances. Drip-dry paints become runny when stirred or applied to a surface but not when held on the brush. Some landslides are caused when normally stable mixtures of clay and water are disturbed.

Exercise

Undertake an investigation to determine how the viscosity of a liquid depends on its temperature. Your investigation should include a **plan**, all recorded **observations**, data **analysis**, a **conclusion** based on your results and an **evaluation** of your methods and findings.

REVIEW QUESTIONS

1 Figure 6.11 shows a free-body force diagram of a sphere falling through a fluid and a velocity–time graph starting from the moment the sphere is released, where U is the upthrust on the sphere, W is its weight and F is the viscous drag.

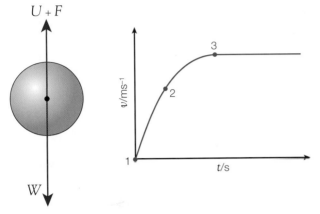

Figure 6.11 ▲

a) The acceleration of the sphere at point 1 on the graph is

A	$0\,\mathrm{m\,s^{-2}}$	B	$2.5\,\mathrm{m\,s^{-2}}$
C	$4.9\,\mathrm{m\,s^{-2}}$	D	$9.8\,\mathrm{m\,s^{-2}}$

b) The acceleration at point 3 is

A	$0\,\mathrm{m\,s^{-2}}$	B	$2.5\,\mathrm{m\,s^{-2}}$
C	$4.9\,\mathrm{m\,s^{-2}}$	D	$9.8\,\mathrm{m\,s^{-2}}$

c) Between points 1 and 2 on the graph

A	$W = F$	B	$W = U$
C	$W = U + F$	D	$W > U + F$

d) At point 3

A	$W = F$	B	$W = U$
C	$W = U + F$	D	$W > U + F$

2 A spherical balloon of radius 20 cm contains 50 g of air.

a) Calculate the density of air in the balloon.

b) Comment on why it differs from that given in Table 6.1 (page 50).

3 The principle of a liquid barometer is to 'balance' the air pressure with the pressure due to a column of the liquid. Calculate the column length of a mercury barometer when the atmospheric pressure is $1.01 \times 10^5\,\mathrm{Pa}$.

4 The density of the water in the Dead Sea is about $1200\,\mathrm{kg\,m^{-3}}$.

a) Calculate the volume of water that needs to be displaced for a person of weight 600 N to float in this sea.

b) How does this compare with the volume that needs to be displaced for the person to float in a swimming pool?

5 Oil is often pumped underground or undersea through pipes that are hundreds of kilometres long.

a) State how the diameter of the pipe, the place of origin and the temperature of the oil affect the rate of flow through the pipe.

b) The oil companies need to transport the oil as quickly as possible. Why does the flow rate need to be restricted?

6 Use Stokes' law to show that the unit of viscosity is $\mathrm{N\,s\,m^{-2}}$.

7 Droplets in a deodorant spray have a mass of about $4 \times 10^{-12}\,\mathrm{kg}$ and a radius of around 0.1 mm.

a) Estimate the terminal velocity of the droplets in air of viscosity $2 \times 10^{-5}\,\mathrm{N\,s\,m^{-2}}$.

b) How does your estimate compare with the speed of falling raindrops?

7 Solid materials

Solid materials play a vital role in our lives. Engineering components, sports equipment and the bones in our bodies have been designed or have evolved to be ideally suited for their purpose.
In this section, you will study the properties of a range of materials and consider their significance in a variety of applications.

A simple example of how confectionery products can be compared is by grading some of their properties on a scale of 1–10. A group of testers can judge how **hard, smooth, chewy, sticky, crunchy, creamy, sweet** or **brittle** each product is, and the results may be analysed to give a taste profile.

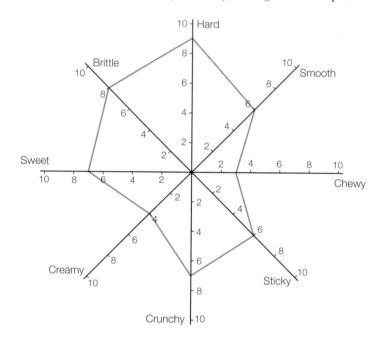

Figure 7.1 ▶
Spider diagram for sweets

Exercise

Use the properties given in Figure 7.1 – or any others that you think could be used – to describe a number of common confectionery products. Find the average grade for each property and construct a spider diagram to display the taste profile.

Most of these properties are subjective and loosely described. In physics and engineering, the properties of solid materials need to be clearly defined and universally accepted.

7.1 Elastic and plastic deformation

A material undergoing **elastic** deformation will return to its original dimensions when the deforming force is removed. A **plastic** material will remain deformed.

A rubber band is often called an 'elastic' band. It can be stretched to several times its normal length and will return to its starting length when the tension is removed. Less noticeably, steel wires or cables can also behave elastically.

Modelling clay is a simple example of materials that deform plastically. It can be pulled, squeezed and twisted into the desired shape.

Some materials can behave in an elastic or plastic manner depending on the nature of the deforming forces. A thin steel sheet will deform elastically when small forces are applied to it, but the huge forces of a hydraulic press will mould the sheet into car panels. Once shaped, the panels regain their elastic properties for everyday stresses.

You will investigate the elastic and plastic properties of materials using force–extension graphs in Section 7.3.

7.2 Properties of solid materials

Hardness

Hardness is a surface phenomenon. The harder the material, the more difficult it is to indent or scratch the surface.

A simple method to compare the hardness of two materials involves finding out which scratches the surface of the other.

Experiment

Comparing the hardness of a range of solids.

Collect a number of blocks of different woods, metals, plastics, etc (for example, from a materials kit) and use a corner of each block to try to score the other blocks. Grade the materials in terms of hardness.

The **Mohs** scale of hardness grades ten minerals from the softest – talc, which is rated as 1 on the scale – to the hardest – diamond, which has a rating of 10. In engineering, the hardness of a metal is measured using the Brinell Hardness Number (BHN), which is the ratio of the load applied to a small steel sphere to the area of the indentation it makes in the surface of the metal being tested.

Stiffness

A **stiff** material exhibits very small deformations even when subjected to large forces. It would require a great force to bend the upright of a laboratory stand even by a few millimetres, whereas a polythene metre rule can be bowed easily. Similarly, a steel piano wire is much stiffer than a rubber band, as it needs a very large force to produce a small extension.

The stiffness of a material is measured in terms of its **modulus of elasticity**. You will learn more about stiffness in Section 7.4.

Toughness

A **tough** material is able to absorb the energy from impacts and shocks without breaking. Tough metals usually undergo considerable plastic deformation in order to absorb the energy.

Car tyres are made from tough rubber/steel compositions. They are designed to withstand impacts from irregular surfaces. The absorbed energy is transferred to the tyres as internal energy – that is, the tyres get hot after a journey.

The energy per unit volume absorbed when materials are deformed will be studied in Section 7.4.

Brittleness

A **brittle** object will shatter or crack when subjected to dynamic shocks or impacts. Brittle materials undergo little or no plastic deformation before breaking.

The glass used for car windscreens is brittle, as are cast iron gratings, concrete and biscuits.

Strength

An object is **strong** if it can withstand a large force before it breaks. The strength of a material will depend on its size – for example, a thick cotton thread requires a bigger breaking force than a thin copper wire.

The strength of a material is therefore defined in terms of its breaking **stress**, where stress is the force per unit cross-sectional area.

Malleability

A **malleable** material can be hammered out into thin sheets. Gold is very malleable and can be hammered into 'gold leaf', which is used to decorate pottery, picture frames, etc.

Ductility

Ductile materials can be drawn into wires. Copper wires are used extensively for electrical connections and are produced by drawing out cylinders to the desired thickness.

Although most ductile materials are also malleable, the reverse is not always true. Many malleable materials will shred or break when extended.

7.3 Hooke's law

You may be familiar with a simple experiment investigating the extension of a spring. The spring is loaded by adding masses to a weight hanger and the extensions are measured as the load increases until the spring is 'uncoiled'. The apparatus for the experiment and a sketch of the resulting load–extension graph are shown in Figure 7.2.

Definition

Hooke's law states that, up to a given load, the extension of a spring is directly proportional to the force applied to the spring:

$F = k\,\Delta x$

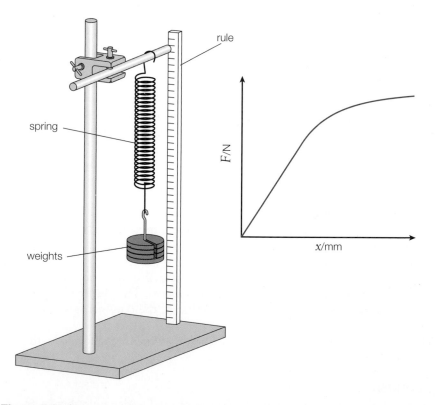

Figure 7.2 ▲

The graph initially has a linear region when the spring is still coiled. As the spring loses its 'springiness,' the extension increases disproportionately.

The constant k represents the 'stiffness' of the spring and is called its **spring constant**.

Worked example

Estimate the spring constant of:

a) a 0–10 N Newton-meter

b) a suspension spring of a family car.

Answer

a) Assume the scale is 10 cm long:

$$k = \frac{F}{\Delta x} = \frac{10\,\text{N}}{0.10\,\text{m}} = 100\,\text{N}\,\text{m}^{-1}$$

b) If the mass of the car is 1600 kg, the force on each of four springs will be $(400\,\text{kg} \times 9.8\,\text{m}\,\text{s}^{-2}) \approx 4000\,\text{N}$.

To take the weight off the spring, the car body needs to be raised about 10 cm:

$$k = \frac{F}{\Delta x} = \frac{4000\,\text{N}}{0.10\,\text{m}} = 4 \times 10^4\,\text{N}\,\text{m}^{-1}$$

The spring experiment is a useful introduction to this section. Most materials behave like springs to some degree, as the bonds between atoms and molecules are stretched when they are loaded. By studying the force–extension graphs of materials, we can see if the material obeys Hooke's law and examine many other properties.

Experiment

Investigating the properties of copper wire

A thin copper wire is clamped between wooden blocks at one end of a bench and passed over a pulley wheel at the opposite end of the bench (Figure 7.3). The wire should be at least two metres long (the longer the wire, the bigger the extension for a given load, which means that more accurate readings can be taken).

A weight hanger is attached to the wire close to the pulley, and a metre rule is fixed to the bench as shown in Figure 7.3. A strip of sticky tape is attached to the wire so that its edge is on the lower end of the metre scale. (The length of the wire from the fixed end to the tape and the diameter of the wire can be taken now, as they may be used at a later stage.)

Weights are placed on the hanger in 200 g increments (2.0 N), and the corresponding extensions are read off the scale. When the copper starts to 'give', a few weights should be removed (tricky) and the wire will continue stretching until it breaks. (It is important that safety glasses are worn to reduce the risk of eye damage when the wire breaks.)

A graph of force against extension is plotted. A typical graph for a three-metre length of 26 swg wire is shown in Figure 7.4.

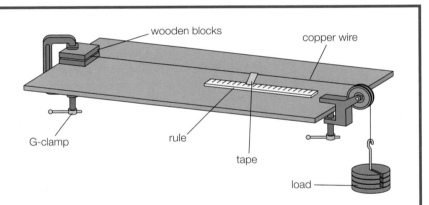

Figure 7.3 ▲ Stretching copper wire

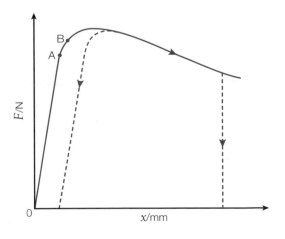

Figure 7.4 ▲
Force–extension curve for copper wire

Force–extension graphs

It is not always possible to use the spring set-up to investigate the tensile properties of materials. Metals, for example, will often require very large forces to produce measurable extensions, and so different arrangements or specialised equipment are needed.

Elastic and plastic behaviour during stretching

Figure 7.4 shows a steep linear region followed by a region of large extension with reducing force. In the initial section (O–A), the extension is proportional to the applied force, so Hooke's law is obeyed.

If the load is removed from the wire up to the limit of proportionality, or even a little beyond this, the wire will return to its original length. This is known as the **elastic** region of the extension, in which loading and unloading are **reversible**. Arrows are drawn on the graph to illustrate the load–unload cycles.

The atoms in a solid are held together by **bonds**. These behave like springs between the particles, and as the copper wire is stretched, the atomic separation increases. In the elastic region, the atoms return to their original positions when the deforming force is removed.

Beyond point B on the graph, the wire ceases to be elastic. Although the wire may shorten when the load is removed, it will not return to its original length – it has passed the point of reversibility and has undergone **permanent deformation**.

As the load is increased, the wire **yields** and will not contract at all when the load is removed. Beyond this **yield point** – point C on the graph – the wire is **plastic** and can be pulled like modelling clay until it breaks. If the broken end of the wire is wound around a pencil, the plasticity can be felt when the wire is extended. In the plastic region, the bonds between the atoms are no longer being stretched and layers of atoms slide over each other with no restorative forces.

A very strange effect is noticed if the load is removed during the plastic phase and the wire is reloaded: the wire regains its springiness and has the same stiffness as before (Figure 7.5).

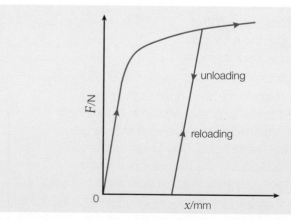

Figure 7.5 ▶
Repeated loading and unloading of copper wire

The ability of some metals to be deformed plastically and then regain their elasticity is extremely important in engineering. A mild steel sheet can be pressed into a mould to the shape of a car panel. After the plastic deformation, the stiffness and elasticity of the steel is regained and further pressings of the panel are also possible.

Steel wires

It is possible to use a similar set-up to the copper wire experiment for steel wires, but as these are much stiffer and yield after relatively small fractional increases in length, it is best not to attempt the investigation without specialist equipment.

Steel is produced by mixing iron with small quantities of carbon. The properties of the steel depend on the percentage of carbon and the heat treatment of the steel. Mild steel contains less than 0.25% carbon and exhibits similar plastic behaviour to copper, whereas high carbon steel is usually quench-hardened (rapidly cooled by dipping into oil or water) and is quite brittle.

Figure 7.6 shows force–extension curves for mild steel and high carbon steel wires of similar length and diameter as the copper wire.

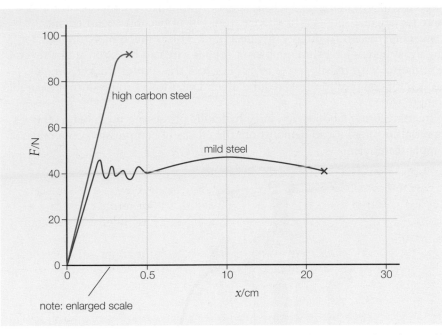

note: enlarged scale

Figure 7.6 ◄
Force–extension curves for high carbon and mild steel

The graph for mild steel shows a similar trend to that for copper, but the yield point is at a lower fractional extension than for copper and the curve is initially quite 'jerky'. Considerable extension is seen after the mild steel yields, but the final break occurs at a much lower percentage increase than for copper. As more carbon is added to the steel, the plastic region is reduced, as shown in the second graph.

Natural rubber

A force–extension graph for a rubber band can be obtained in a similar manner to the experiment with the spring, but an alternative method is shown in Figure 7.7.

It is best to use a short thin rubber band, as it will stretch to several times its original length and thick bands are difficult to break.

To alter the length of the band, the boss is loosened and moved up the stand. The force at each extension is read off the newton meter.

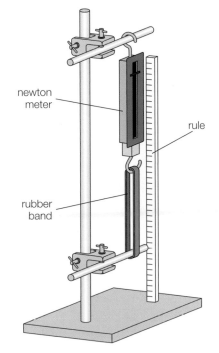

Figure 7.7 ▲
Stretching a rubber band

Figure 7.8 ◄
Force–extension curve for rubber band

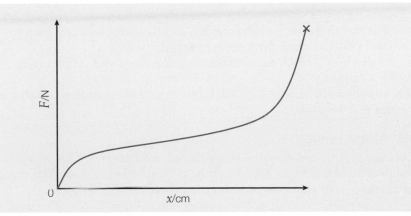

The rubber band stretches very easily at first, reaching a length of three or four times its original value. It then becomes very stiff and difficult to stretch as it approaches its breaking point.

It should be noted that the band returns to its original length if the force is removed at any stage prior to breaking – that is, the rubber band is elastic.

Force–compression graphs

Up to now we have considered only the behaviour of materials that have been subjected to stretching – or **tensile** forces. If weights were placed on top of a large rectangular sponge, the sponge would be noticeably squashed and force–compression readings easily measured. This is much more difficult for metals. Whereas long thin samples of copper wire can be extended by several centimetres with tensile forces of 50 N or less, the shorter thicker samples needed for compression tests require much larger forces to produce measurable compressions.

In engineering laboratories, large hydraulic presses are used, but a school's materials testing kit can demonstrate the effects of compression on a range of sample materials.

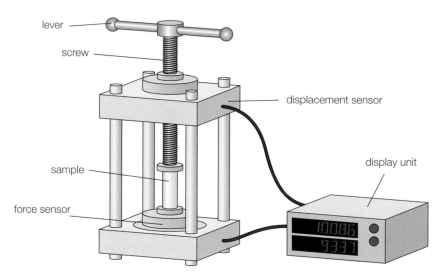

Figure 7.9 ▲
Compression testing kit

The sample is placed into the press and the screw is tightened to hold it firmly in position. Force and position sensors are connected to a display module or to a computer via a data processing interface, and the readings are zeroed.

The sample is compressed by rotating the lever clockwise. A series of values of force and the corresponding compression is taken and a force–compression graph drawn.

The elastic region is similar to that seen in the tensile tests. The bonds are 'squashed' as the atoms are pushed together, and the particles move back to their original position when the force is removed.

Plastic behaviour is much more difficult to examine as the samples twist and buckle at failure.

Some kits allow the sample to be stretched as well as compressed, so that a comparison may be made.

Elastic strain energy

The concept of potential energy was introduced in Chapter 5. This relates to the ability of an object to do work by virtue of its position or state. The elastic potential energy – or elastic strain energy – is therefore the ability of a deformed material to do work as it regains its original dimensions. The work done stretching the rubber of a catapult (slingshot) is transferred to elastic strain energy in the rubber and then to kinetic energy of the missile on its release.

The work done during the stretching process is equal to the average force times the distance moved in the direction of the force:

$$\Delta W = F_{ave} \Delta x$$

The work done on a wire, and hence its elastic strain energy, can be obtained from a force–extension graph.

For the Hooke's law region of the graph (O–A) in Figure 7.10, the average force is $\dfrac{F_{max}}{2}$, so the work done is:

$$\Delta W = \tfrac{1}{2} F_{max} \Delta x$$

This represents the area between the line and the extension axis – that is, the area of the triangle made by the line and the axis. Similarly, the work done when the force is constant (A–B on the graph) will be the area of the rectangle below the line.

For any force–extension graph, the elastic strain energy is equal to the area under the graph. To calculate the energy for non-linear graphs, the work equivalent of each square is calculated and the number of squares beneath the line is counted. For estimated values, the shape can be divided into approximate triangular or rectangular regions.

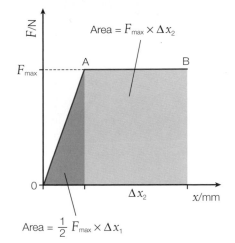

Figure 7.10 ▲
Elastic strain energy

Worked example

Estimate the elastic strain energy in the material that produces the force–extension graph in Figure 7.11.

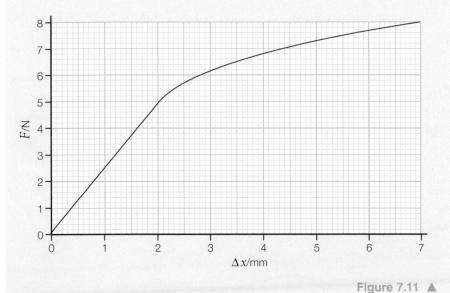

Figure 7.11 ▲

Answer

Area of 'Hooke's law region' $= \tfrac{1}{2} (5\,\text{N}) (2 \times 10^{-3}\,\text{m}) = 5 \times 10^{-3}\,\text{J}$

Area of trapezium $= \tfrac{1}{2} (5\,\text{N} + 8\,\text{N}) (5 \times 10^{-3}\,\text{m}) = 4 \times 10^{-2}\,\text{J}$

Elastic strain energy $= 5 \times 10^{-2}\,\text{J}$

For a material that is extended within the Hooke's law limit, the strain energy can be calculated in terms of the spring constant of the material:

$$\Delta W = \tfrac{1}{2} F \Delta x \text{ and } F = k \Delta x$$

$$\Delta W = \tfrac{1}{2} (k \Delta x) \Delta x = \tfrac{1}{2} k \Delta x^2$$

Definition

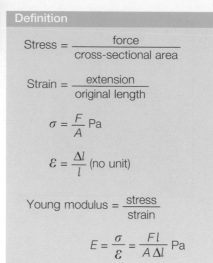

Stress = $\dfrac{\text{force}}{\text{cross-sectional area}}$

Strain = $\dfrac{\text{extension}}{\text{original length}}$

$\sigma = \dfrac{F}{A}$ Pa

$\varepsilon = \dfrac{\Delta l}{l}$ (no unit)

Young modulus = $\dfrac{\text{stress}}{\text{strain}}$

$E = \dfrac{\sigma}{\varepsilon} = \dfrac{Fl}{A\,\Delta l}$ Pa

7.4 Stress and strain: the Young modulus

In everyday terms, one might say that students put their teacher under stress and the teacher takes the strain. Likewise, when a stress is applied to a material, the strain is the effect of that stress.

A property of materials that undergo tensile or compressive stress is the **Young modulus**, E.

Worked example

A steel wire of length 2.00 m and diameter 0.40 mm is extended by 4.0 mm when a stretching force of 50 N is applied. Calculate:

1 the applied stress
2 the strain on the wire
3 the Young modulus of steel.

Answer

1 Stress, $\sigma = \dfrac{F}{A} = \dfrac{50\,\text{N}}{\pi(0.20 \times 10^{-3}\,\text{m})^2} = 4.0 \times 10^8\,\text{Pa}$

2 Strain, $\varepsilon = \dfrac{\Delta l}{l} = \dfrac{4.0 \times 10^{-3}\,\text{m}}{2.00\,\text{m}} = 2.0 \times 10^{-3}$ (0.20%)

3 Young modulus, $E = \dfrac{\sigma}{\varepsilon} = \dfrac{4.0 \times 10^8\,\text{Pa}}{2.0 \times 10^{-3}} = 2.0 \times 10^{11}\,\text{Pa}$

The importance of stress and strain as opposed to force and extension is that they are properties of the material: a stress–strain graph is always the same for a given material, whereas a force–extension graph depends on the dimensions of the sample used.

For most investigations, the cross-sectional area of the sample is a few square millimetres at most (1 mm² = 1×10^{-6} m²), so the stress is very large (up to several hundred MPa). As strains are often very small, the Young modulus can be hundreds of GPa for metals.

Stress–strain graphs

The shape of stress–strain graphs is much the same as that of force–extension graphs, although for large extensions a reduction in cross-sectional area will result in an increase in the stress for a particular load. The main advantage is that the information gained from the graph relates to the properties of the material used and not just those of the particular sample tested.

Figure 7.12 shows a stress–strain graph for the copper wire used earlier in the stretching experiment. It should be noted that the strain scale is expanded over the initial 1%.

Figure 7.12 ▶
Stress–strain graph for copper wire

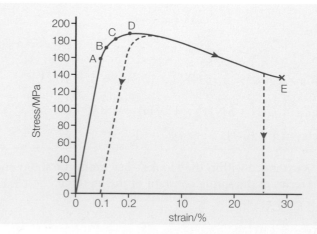

- O–A represents the **Hooke's law region**. Strain is proportional to stress up to this point. The **Young modulus** of copper can be found directly by taking the gradient of the graph in this section.
- B is the **elastic limit**. If the stress is removed below this value, the wire returns to its original state.
- The stress at C is termed the **yield stress**. For stresses greater than this, copper will become **ductile** and deform plastically.
- D is the maximum stress that the copper can endure. It is called the **ultimate tensile strength** (**UTS**) or simply the strength of copper.
- E is the breaking point. There may be an increase in stress at this point due to a narrowing of the wire at the position on the wire where it breaks, which reduces the area at that point.

Exercise

Use your values of length and diameter from the stretching copper wire experiment (page 61) to calculate the stress and strain for each reading of force and extension.

Plot a graph of stress against strain and use the graph to determine a value of the Young modulus, the yield stress and the UTS of copper.

If you do not have any readings, use the following typical set of results from a similar experiment:

Length of wire = 3.00 m

Mean diameter of wire = 0.52 mm

Complete Table 7.1 and use the stress–strain graph (for extensions up to 300 mm only) to determine the Young modulus, the yield stress and the UTS of the copper.

Force /N	Extension /mm	Stress /Pa	Strain /%
0	0		
5.0	0.5		
10.0	1.0		
15.0	1.5		
20.0	2.0		
25.0	2.5		
30.0	3.0		
35.0	4.0		
40.0	6.0		
35.0	300		
30.0	750		

Table 7.1 ▲

Figure 7.13 provides stress–strain graphs to show the comparative properties of high carbon steel, mild steel and copper. The early part of the strain axis is extended to show the Hooke's law region more clearly.

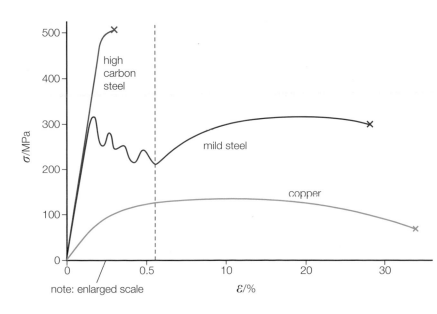

note: enlarged scale

Figure 7.13 ◄
Stress–strain curves for metals

The graphs illustrate the different behaviour of the three materials. The gradient of the Hooke's law region is the same for the steels and is greater than that for copper. The steels have a Young modulus of about 200 GPa and are stiffer than copper ($E \approx 130$ GPa).

High carbon steel is the strongest as it has the greatest breaking stress (UTS), but it fractures with very little plastic deformation – it is brittle. Quench-hardened high carbon steel is commonly used for cutting tools and drill bits.

Mild steel has an upper and lower yield point and a fairly long plastic region. Because it regains its high stiffness when the deforming force is removed, it is ideal for pressing into car body parts.

The long plastic region of copper means that it is very ductile and so is easily drawn into wires. This, together with its low resistivity (see Section 14), makes copper an invaluable material for the electrical industry.

Compressive stress–strain graphs
If the initial length and cross-section of the sample used for the force–compression graph are measured, compressive stresses and strains can be calculated. The Young modulus of the material under compression can then be found by taking the gradient of the stress–strain graph. It is generally the same as the modulus found in tensile tests.

For artificial joints used in hip replacement surgery, it is important that the Young modulus of the artificial joint, as well as its hardness and durability, is similar to that of the bone it replaced, otherwise differences in compression for similar stresses could lead to deterioration of the bone around the artificial joint.

Worked example

Estimate the stress on the lower leg bones of a stationary erect human being. Explain why your value is likely to be much less than the ultimate compressive stress of bone.

Answer
Assume that the person has a mass of 60 kg and that the supporting bones are 3.0 cm in diameter.

Force on each bone $F = 30$ kg $\times$ 10 m s^{-2} = 300 N

Cross-sectional area $A = \pi (1.5 \times 10^{-2}$ m$)^2 = 7.1 \times 10^{-4}$ m^2

Stress $= \dfrac{F}{A} \approx 400$ kPa

When walking, running and jumping, the stresses on the bones will be much bigger.

Energy density
The energy density is the work done in stretching a specimen (the strain energy stored) per unit volume of the sample.
For a wire that obeys Hooke's law:

$$\text{energy density} = \frac{\text{work done}}{\text{volume}} = \frac{\Delta W}{A\,l} = \frac{F_{\text{ave}}\,\Delta l}{A\,l} = \tfrac{1}{2}\frac{F\,\Delta l}{A\,l}$$

$$= \tfrac{1}{2}\frac{F}{A} \times \frac{\Delta l}{l}$$

$$= \tfrac{1}{2}\ \text{stress} \times \text{strain}$$

This is represented on the stress–strain graph by the area between the line and the strain axis.

As with the force–extension graphs and energy stored, the area under any stress–strain graph represents the energy density.

The ability of a specimen to absorb a large amount of energy per unit volume before fracture is a measure of the **toughness** of the material. Mild steel and copper with 20–30% plastic strains absorb large amounts of energy before breaking and so are tough materials. High carbon steel fractures with little plastic deformation and the area under the stress–strain graph is small, so the material has a low energy density and is **brittle**.

Car tyres need to absorb energy as they roll over uneven surfaces and so are very tough, whereas glass and ceramic materials show very little plastic extension and are brittle.

Hysteresis in rubber

The elastic property of rubber is complex. When a rubber band is stretched and relaxed it does return to its original length, but the manner by which it does so is very different to that of a metal.

The experiment carried out to obtain a force–extension graph for a rubber band can be changed so that the band is loaded as before, but this time the force is removed a little at a time before the rubber band breaks, until the force is zero and the original length of the rubber band is regained.

Figure 7.14 shows such a variation of stress for a rubber band that is extended to four times its length before it is relaxed.

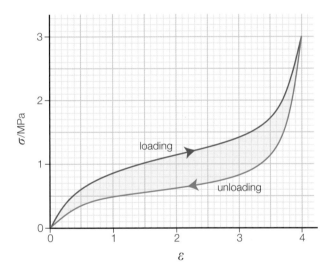

Figure 7.14 ◄
Hysteresis loop for rubber

The arrows indicate the loading and unloading of the rubber band. The area under the loading curve represents the work done per unit volume **on** the band as it stretches, and the area beneath the unloading line is the work done per unit volume of the band **by** the band as it relaxes. This difference is known as **hysteresis**, and the shaded area enclosed by the loading and unloading curves is called a hysteresis loop.

The hysteresis loop for a stress–strain graph represents the energy per unit volume transferred to internal energy during the load–unload cycle.

The force–extension graph for rubber is identical in shape to the stress–strain graph, but the area of the hysteresis loop now represents the total energy transferred to internal energy for each cycle.

If a rubber band is repeatedly stretched and relaxed in a short time, it will become warm. A simple experiment to demonstrate this effect involves rapidly stretching and relaxing a rubber band several times and then placing it against your lips.

Much of the kinetic energy transferred by a moving vehicle occurs during the hysteresis loops in the tyres, which become quite warm after a journey.

Worked example

Estimate the energy per unit volume transferred to internal energy for each load–unload cycle of the rubber band used for the stress–strain graph in Figure 7.14.

Answer

About six squares are enclosed by the loop.

Each square is equivalent to $0.5\,\text{MPa} \times 0.5 = 0.25\,\text{MJ m}^{-3}$

$\Rightarrow$ the shaded area $\equiv 1.5\,\text{MJ m}^{-3}$

A more accurate value can be obtained by counting small squares. As examination questions usually require an estimated value, such a time-consuming method is inappropriate during a timed test.

Summary of the properties of solid materials

In this section, several properties that describe the behaviour of solid materials have been introduced. Table 7.2 lists these, together with definitions and examples where appropriate.

Property	Definition	Example	Opposite	Definition	Example
Strong	High breaking stress	Steel	Weak	Low breaking stress	Expanded polystyrene
Stiff	Gradient of a force–extension graph. High Young modulus	Steel	Flexible	Low Young modulus	Natural rubber
Tough	High energy density up to fracture: metal that has a large plastic region	Mild steel, copper, rubber tyres	Brittle	Little or no plastic deformation before fracture	Glass, ceramics
Elastic	Regains original dimensions when the deforming force is removed	Steel in Hooke's law region, rubber	Plastic	Extends extensively and irreversibly for a small increase in stress beyond the yield point	Copper, modelling clay
Hard	Difficult to indent the surface	Diamond	Soft	Surface easily indented/scratched	Foam rubber, balsa wood
Ductile	Can be readily drawn into wires	Copper	Hard, brittle		
Malleable	Can be hammered into thin sheets	Gold	Hard, brittle		

Table 7.2 ▲

Worked example

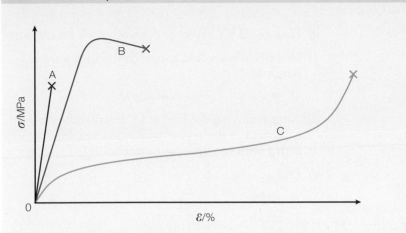

Figure 7.15 ▲ Stress–strain curves

Graphs A, B and C in Figure 7.15 represent the variations of strain with stress for three different materials.

1 Grade the samples in terms of:

   a) strength
   b) stiffness
   c) toughness.

2 Suggest a possible material for each of A, B and C.

Answer

1 **a)** B, C, A
   b) A, B, C
   c) C, B, A

2 A: cast iron, glass or ceramic
   B: mild steel, brass
   C: rubber or other similar polymer

REVIEW QUESTIONS

1 Figure 7.16 shows the stress–strain graphs for four different materials.

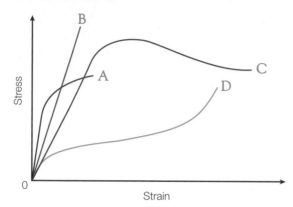

Figure 7.16 ▲

a) Which material is toughest?

A   B   C   D

b) Which material is strongest?

A   B   C   D

c) Which material is a polymer?

A   B   C   D

d) Which material is brittle?

A   B   C   D

2 a) State Hooke's law.

b) A spring of length 10.0 cm extends to 12.5 cm when an 8.0 N weight is suspended from it. Calculate the spring constant of the spring.

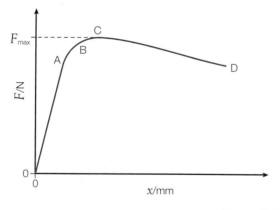

Figure 7.17 ▲

3 a) Identify from Figure 7.17:

   i) the yield point

   ii) the Hooke's law region

   iii) the plastic region on the force–extension curve for copper.

b) How could you estimate the work done on the wire?

c) Describe what would happen if the load were removed:

   i) at A   ii) at D

d) After removing the load at D, describe the behaviour of the copper wire when weights are added until the force again reaches F_{max}.

4 a) Define:

   i) stress

   ii) strain

   iii) Young modulus.

b) Calculate the Young modulus of a wire of diameter 0.60 mm and length 2.00 m if a load of 50 N extends the wire by 2.5 mm.

5 Metals may be ductile and malleable. Explain these terms, and state two examples of metals that usually possess such properties.

6 The stress–strain graph in Figure 7.18 shows the hysteresis curve for a rubber band.

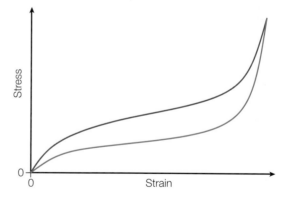

Figure 7.18 ▲
Hysteresis curve for rubber band

a) What is meant by the term hysteresis?

b) Which line represents the loading curve?

c) What does the area within the loop represent?

d) Describe the changes in the molecular structure of the rubber during one loading–unloading cycle.

Unit 1 test

Time allowed: 1 hour 20 minutes
Answer **all** of the questions.

For Questions 1–4, select one answer from A–D.

1 Which of the following quantities is **not** a vector?

  A Acceleration

  B Displacement

  C Kinetic energy

  D Velocity   **[1]**

                 [Total: 1 mark]

2 A 10 kg mass suspended by a steel wire of cross-sectional area 2.0 mm² produces a stress on the wire of approximately:

  A 50 Pa

  B 50 kPa

  C 5 MPa

  D 50 MPa   **[1]**

                 [Total: 1 mark]

3 An object will always continue to move with a constant velocity if:

  A a uniform resultant force is acting on it

  B all the forces acting on it are in equilibrium

  C it is in a vacuum

  D two forces of equal magnitude act on it.   **[1]**

                 [Total: 1 mark]

4 The resistive force on a ball-bearing falling through oil does **not** depend on:

  A its diameter

  B its mass

  C its velocity

  D the viscosity of the oil.   **[1]**

                 [Total: 1 mark]

For Questions 5 and 6, choose the appropriate letter from the list to indicate which value best completes the sentence. Each answer may be used once, more than once or not at all.

  A 0 m s⁻²

  B between 0 and 9.8 m s⁻²

  C 9.8 m s⁻²

  D greater than 9.8

5 The acceleration of a parachutist just prior to landing is:

  A

  B

  C

  D   **[1]**

                 [Total: 1 mark]

6 The vertical acceleration of a high jumper at the top of her leap is:

  A

  B

  C

  D   **[1]**

                 [Total: 1 mark]

Questions 7 and 8 relate to the stress–strain graph in Figure 1, which is for a metal wire stretched until it breaks. Use the graph to determine which answer best completes the statement.

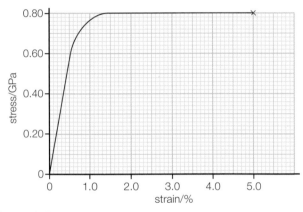

Figure 1 ▲

7 The Young modulus of the metal is:

  A 16 GPa

  B 53 GPa

  C 74 GPa

  D 120 GPa   **[1]**

                 [Total: 1 mark]

8 The wire is must likely to be undergoing plastic deformation where the strain is in the region:

  A 0–0.5%

  B 0.5–1.5%

  C 0.5–5.0%

  D 1.0–5.0%   **[1]**

                 [Total: 1 mark]

For Questions 9 and 10, which of the graphs in Figure 2 best represents the situation described? Each graph may be used once, more than once or not at all.

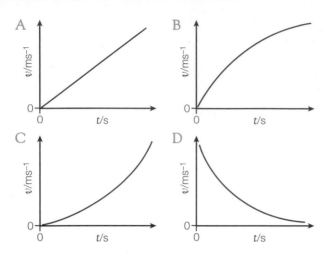

Figure 2 ▲

9 The motion of a rocket carrying a space exploration capsule immediately after lift off.

A

B

C

D   [1]

[Total: 1 mark]

10 The descent of a sky diver just after leaving the aeroplane.

A

B

C

D   [1]

[Total: 1 mark]

11 Copy and complete the gaps in the passage by selecting appropriate words from the following list.

elastic   friction   hard   kinetic energy
plastic   polymers   stiff   strong

A golf club has a graphite shaft with a rubber grip at one end and a stainless steel head at the other. The rubber and the graphite are _____ with long-chain molecular structures. The shaft is described by the manufacturer as _____ because it has a high Young modulus, and the _____ between the soft rubber grip and the golfer's hands reduces the chances of the club slipping on impact with the ball.

The face of the club-head needs to be _____ to protect the surface from being scratched by stones, etc. When a golf ball is struck by the club-head it undergoes _____

deformation and so rapidly regains its original shape. It moves off at high speed because most of the _____ of the club-head is transferred to the ball.   [3]

[Total: 3 marks]

12 The cat in Figure 3 is sitting on a cushion.

Figure 3 ▲

a) Identify two pairs of Newton's third law forces that relate to the cat. State the type of force involved for each pair and the directions in which they act.   [4]

b) What condition is necessary for the cat to be in equilibrium?   [2]

[Total: 6 marks]

13 A child on a swing is pulled back with a horizontal force, F, so that the chains are at an angle of 30° to the vertical. At that instant, the child is in equilibrium.

Figure 4 ▲

a) Draw a free-body force diagram for the child–seat combination, labelling the combined weight of the child and seat, W, the tension in the chains, T, and the holding force, F.   [3]

b) Give expressions for the vertical and horizontal components of T.   [1]

c) If the combined weight of the child and seat is 250 N, show that F needs to be about 140 N to maintain equilibrium.   [3]

[Total: 7 marks]

1

14 A tennis player returns a shot by striking the ball just above ground level. The ball leaves the racquet with a horizontal velocity component of $20.0\,\text{m s}^{-1}$ and a vertical velocity component of $5.0\,\text{m s}^{-1}$.

   a) Calculate the maximum height reached by the ball and the time taken to reach this height. **[2]**

   b) The tennis court is about 24 metres (78 feet) long and 11 metres (36 feet) wide and the maximum height of the net is 1.1 m (42 inches). Use your answers to part **a)** to discuss whether it is possible for the ball to clear the net and land within the confines of the court. **[2]**

                                    [Total: 4 marks]

15 A ski-lift transports 50 skiers to the start of a run that is at an altitude of 100 m above its base.

   a) If the average mass of a skier plus equipment is 60 kg and the journey takes five minutes, show that the average power generated by the motor is about 10 kW. **[2]**

   b) If the electrical power input during the trip is 12 kW, calculate the efficiency of the motor. **[1]**

                                      [Total: 3 marks]

16 Figure 5 shows the movement of air relative to an aircraft wing. The lines represent the motion of layers of air particles close to the wing.

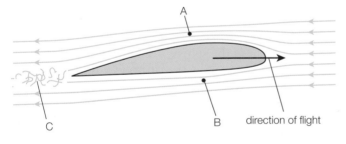

Figure 5 ▲

   At A and B, the layers of air do not cross over each other, but the air at C swirls around and the particles mix together.

   a) Write down a word that describes the flow of air at A and B and another that represents the movement at C. **[2]**

   b) The uplift on an aeroplane is a consequence of the faster moving air at A having a lower pressure than that at B. For a Boeing 737 cruising at $270\,\text{m s}^{-1}$ (Mach 0.8), this pressure difference is about $4 \times 10^3\,\text{Pa}$. Calculate the uplift if the underwing area is $120\,\text{m}^2$. **[2]**

   c) Estimate the mass of the plane and its load. **[1]**

                                      [Total: 5 marks]

17 Figure 6 represents the force–extension curve for a rubber band as it is loaded and then unloaded.

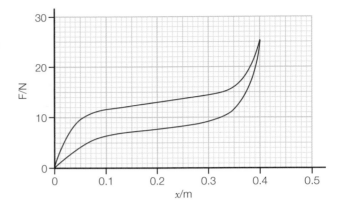

Figure 6 ▲

   a) Sketch and label the graph to show the loading and unloading curves. **[1]**

   b) What is represented by the area enclosed within the loop? Estimate this value from the graph. **[3]**

   c) What physical change in the rubber would occur if it were rapidly stretched and released several times? **[1]**

                                      [Total: 5 marks]

18 A student was asked to describe an experiment to determine the acceleration due to gravity. As an answer the student wrote the following:

   'A steel ball is held by an electromagnet above a trapdoor switch. When the magnet is switched off an electronic clock is switched on and the ball starts to fall. The ball hits the trapdoor and the clock stops.

   The experiment is repeated and a graph is drawn. The gradient of the graph is equal to g.'

   a) Discuss the student's answer, highlighting any incorrect or missing physics. **[4]**

   b) Calculate the height of the ball above the trapdoor if the time for it to fall was 0.60 s. **[2]**

                                      [Total: 6 marks]

19 A clown jumps from a blazing castle onto a strategically placed trampoline.

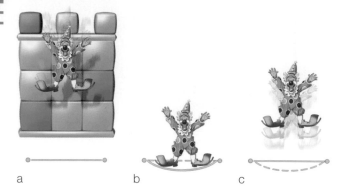

a   b   c

Figure 7 ▲

Describe the energy changes that take place:

a) as the clown is falling towards the trampoline (Figure 7a) [2]

b) from the moment the clown makes contact with the trampoline to the instant he is stationary at the lowest point (Figure 7b) [2]

c) during the time he rises from his stationary position until he reaches his highest position (Figure 7c). [2]

d) Explain why the clown will not bounce back up to the fire. [1]

[Total: 7 marks]

20 A car pulls out of a side road and its engine stalls, leaving it stationary and blocking the carriageway. The driver of a second car, which is on the main road, applies the brakes but cannot prevent a collision with the stationary vehicle.

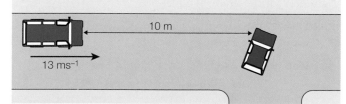

Figure 8 ▲

a) The second car of mass 1500 kg was travelling at 13 m s^{-1} and was 10 m from the stranded car when the brakes were applied. If the average decelerating force between the road and the tyres was 7.2 m s^{-2}, calculate the velocity of the car at the moment of impact. [2]

b) The collision time (from the time of impact until the cars stopped moving) was 0.5 s. Estimate the average force exerted by the moving car onto the side panel of the static vehicle. [3]

c) The Highway Code states that the braking distance for a car travelling at 30 mph (13.4 m s^{-1}) is 14 m. Give one reason why this may be inaccurate. [1]

d) The front end of the moving car has a 'crumple zone', and the doors of the other car are fitted with a 'side impact protection system' (SIPS). Describe how these design features reduce the force of impact between colliding vehicles. [3]

e) A car body repairer is able to beat out the panel back to its original shape. The panels are made from low carbon mild steel. Describe the properties of this steel that enable it to be reshaped after the indentation. [2]

[Total: 11 marks]

21 Figure 9 shows a sphere of weight, W, and radius, r, falling through a liquid.

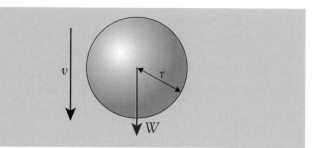

Figure 9 ▲

a) Copy and add labelled arrows to the diagram to represent the upthrust, U, and the viscous drag, F. [2]

b) The radius of the sphere is 1.5 mm and its density is 7800 kg m^{-3}. Show that its weight is about 1.1×10^{-3} N. [2]

c) What condition is necessary for the sphere to reach its terminal velocity? [1]

d) If the upthrust on the sphere is 0.20×10^{-3} N, calculate the value of the terminal velocity of the sphere in a liquid of viscosity 0.35 Pa s. [3]

e) In an investigation into the effect of the temperature on the viscosity of oil, the terminal velocity of a ball-bearing is measured as it falls through the oil over a range of temperatures. Figure 10 shows the apparatus used and a graph of the terminal velocity against the temperature of the oil.

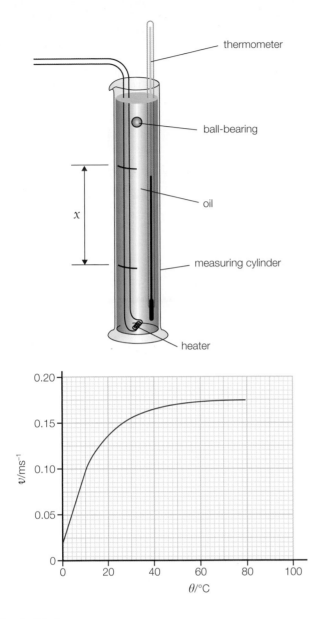

Figure 10 ▲

   i) Use the graph to find the value of the terminal
      velocity of the ball at 30 °C.   **[1]**

  ii) Use the graph to describe how the viscosity of
      the oil changes between 10 °C and 80 °C.   **[3]**

 iii) Some motor oils are termed 'viscostatic'.
      What property of the oil is described by this
      statement?   **[1]**

[Total: 13 marks]
[TOTAL: 80 marks]

Unit 2

Topic 4 Waves

8 Nature of waves

9 Transmission and reflection of waves

10 Superposition of waves

Topic 5 DC electricity

11 Charge and current

12 Potential difference, electromotive force and power

13 Current–potential difference relationships

14 Resistance and resistivity

15 Electric circuits

Topic 6 Nature of light

16 Nature of light

8 Nature of waves

Waves occur in a variety of forms. In communications, sound waves and radio waves carry information. Energy from the Sun reaches us via light, infrared and ultraviolet waves. Other examples are shock waves from earthquakes, water waves, X-rays and γ rays.

The above examples are often very different in their behaviours, but they all have characteristic properties that define them as waves. In this chapter, you will study the nature of different waves. Other properties and applications will be scrutinised more closely in the following two chapters.

8.1 Mechanical oscillations and waves

Mechanical waves

Mechanical waves require a medium for transmission and are generated by vibrating sources. The energy from the vibrations is transferred to the medium and is transmitted by the particles within it.

The passage of a wave through a medium can be demonstrated using a 'slinky' spring. The vibrations can be applied to the spring either parallel to the direction of transmission of the wave or perpendicular to it, as shown in Figure 8.1. The movement of the 'particles' can be observed by marking one loop of the slinky with a coloured pen. You will see that the marked loop moves back and forth or side to side about its initial position.

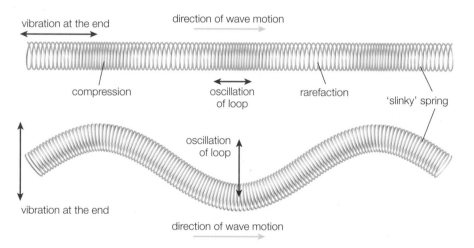

Figure 8.1 ▲

Waves with the particles oscillating parallel to the wave motion are called **longitudinal** waves. Those with oscillations perpendicular to the direction of progression are called **transverse** waves.

Sound is transmitted using longitudinal waves. This can be demonstrated using a lighted candle placed close to a large loudspeaker connected to a signal generator, as shown in Figure 8.2. If the frequency of the signal generator is set to 50 Hz or less, the candle flame will be seen to flicker back and forth by the oscillating air molecules. It should be noted that the bulk of the air does not move with the wave, as the particles vibrate about a fixed mean position.

Close inspection of the longitudinal wave in the slinky in Figure 8.1 shows the movement of tightly packed coils followed by widely spaced sections. These are called **compressions** and **rarefactions**. For sound waves in a gas, they create high and low pressure regions (Figure 8.3).

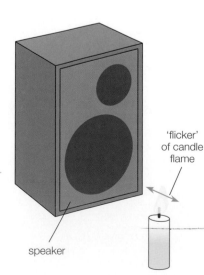

speaker

'flicker' of candle flame

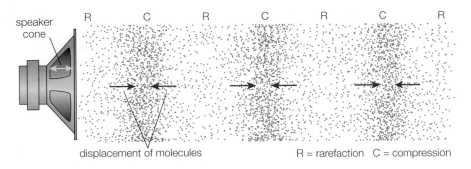

speaker cone

displacement of molecules

R = rarefaction   C = compression

Figure 8.3 ◄
Sound waves in air

If you throw a stone into a still pond you will see waves on the surface spreading outwards in a circular fashion. An object floating on the water will bob up and down as the wave passes, but does not move along the wave (Figure 8.4). The water wave is an example of a transverse wave.

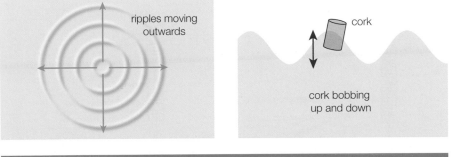

ripples moving outwards

cork

cork bobbing up and down

Figure 8.4 ◄

Exercise

To envisage the motion of the particles in longitudinal and transverse waves is quite difficult using only static diagrams. Many animations that give a much clearer picture of the waves are available. Use a search engine to find 'wave animations' and compare several simulations. Make a note of the site that you believe gives the best animation.

Oscillations

You have seen that mechanical waves are produced by vibrations. A detailed study of oscillations is required in Unit 5 of the A2 course, but a brief outline is useful to explain the particle behaviour in waves.

A simple pendulum, a mass on a spring and a rule clamped on a bench are familiar examples of oscillating systems (Figure 8.5). In all cases, the motion is repetitive about a fixed point, with the object at rest at the extremes of the motion and moving at maximum speed in either direction at the midpoint.

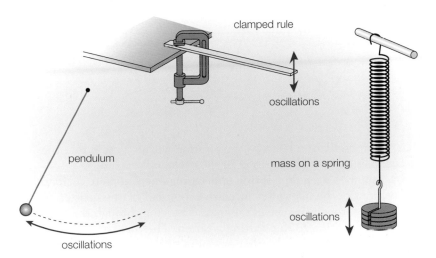

clamped rule

oscillations

pendulum

mass on a spring

oscillations

oscillations

Figure 8.5 ◄
Oscillating systems

Amplitude, **A**, is the maximum displacement from the mean position (metres, m).

Period, **T**, is the time taken for one complete oscillation (seconds, s).

Frequency, **f**, is the number of complete oscillations per second (hertz, Hz).

Three properties can be used to describe an oscillation. They are: the amplitude, A, the period, T, and the frequency, f.

If an oscillator has a frequency of 10 Hz, it will complete ten oscillations every second, so it follows that each oscillation will take 0.1 s. The relationship between the frequency and the period can therefore be written as:

$$f = \frac{1}{T}$$

Displacement–time graphs for oscillating particles

If a soft marker pen is attached to a mass on a spring so that it moves up and down with the mass and a piece of white card is moved at constant speed across the tip of the pen, a trace of the position of the mass over a period of time will be drawn onto the card (Figure 8.6). A more accurate method of plotting the displacement against time can be achieved using a motion sensor with its associated software.

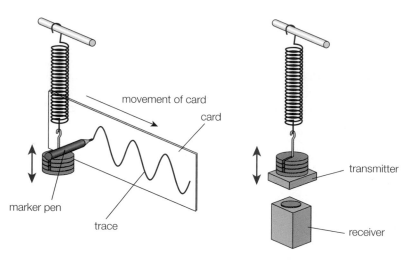

Figure 8.6 ▲

The graph in Figure 8.7 shows that the amplitude of the oscillation is 2.0 cm and the period is 0.8 s. The shape of the repetitive graph is the same as that of a graph of $\sin \theta$ against θ, so the motion is termed sinusoidal. This variation of displacement with time of the particles within a wave is referred to as a **waveform**.

Figure 8.7 ▶

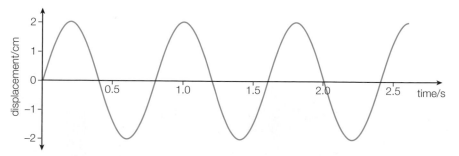

Phase

The phase of an oscillation refers to the position within a cycle that the particle occupies relative to the onset of the cycle.

Figure 8.8 shows a mass on a spring undergoing one complete oscillation and the positions of the mass after successive intervals of one quarter of a period. Halfway through the cycle, the mass is moving through the midpoint in the opposite direction to the starting position. The oscillations at these points are out of step with each other or 'out of **phase**'. Any two oscillations that are half a cycle out of step are said to be in **antiphase**. At the end of the cycle, the mass is moving through the midpoint in the same direction as at the start and the vibrations are now back in phase.

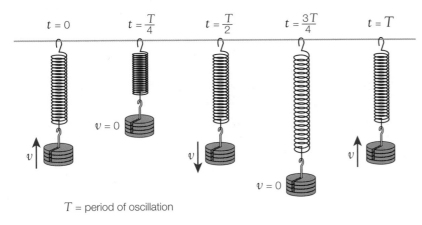

T = period of oscillation

Figure 8.8 ▲ Phases of an oscillation

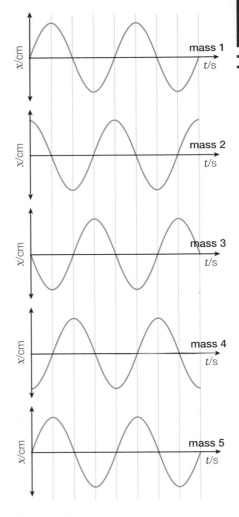

Figure 8.9 ▲

If you consider the oscillations in Figure 8.8 to be five separate masses, you will note that masses 1 and 5 are always in phase, mass 3 is in antiphase with both 1 and 5 and masses 2 and 4 are one quarter of a cycle and three quarters of a cycle out of phase with mass 1.

For a sinusoidal waveform, one cycle represents 2π radians (360°), so oscillations that are in antiphase are said to have a **phase difference** of π radians (180°), while those that differ by one quarter and three quarters of a cycle are $\frac{\pi}{2}$ radians (90°) and $\frac{3\pi}{2}$ radians (270°) out of phase.

The phase relationships of masses 1–5 are represented on the displacement–time graphs in Figure 8.9.

Displacement–distance graphs

Imagine a water wave travelling along the surface of a glass-sided aquarium. A snapshot of the edge of the wave viewed through the glass would reveal a **wave profile** showing the shape of the surface at that instant. This profile shows the positions of the particles along the surface at that time and can be represented on a displacement–distance graph (see Figure 8.10).

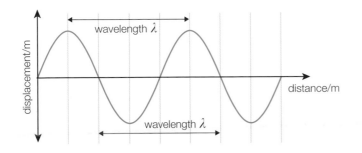

Figure 8.10 ▲ Displacement–distance graph

Although this looks similar to the displacement–time graph, it differs in that it represents the positions of all the particles along that section of the wave as opposed to the motion of a single particle within the wave.

The vibrations of the particles along the wave are all a little out of phase with their neighbours. A Mexican wave at a sports ground illustrates this quite well. When the person to your right starts standing up, you follow a moment later. The person to your left repeats the motion after you, and so the wave moves around the stadium. After a certain length of the wave, the particles come back into phase – for example, the particles at the 'crest' of the wave are all at the top of their oscillation and those in the 'trough' are all at the lower amplitude. The distance between two adjacent positions that are in phase is the **wavelength**, λ, of the wave.

For longitudinal waves such as sound, the displacement of the particles is parallel to the motion, so a wave profile is more difficult to draw, but a displacement–position graph identical to that for transverse waves can be drawn with positive displacements to the right of the mean positions and negative displacements to the left (Figure 8.11).

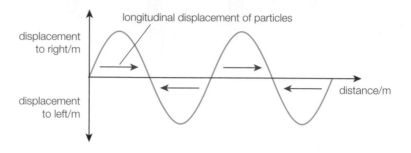

Figure 8.11 ▶

8.2 Electromagnetic waves

Mechanical waves are initiated by vibrating objects passing on some of their energy to the atoms or molecules of a material medium. Electromagnetic waves are created when charged particles are accelerated.

Radio waves are an example of electromagnetic radiation. Electrons in an aerial are made to oscillate using electronic circuitry, and this produces a wave of continuously varying electric and magnetic fields with the same frequency as the oscillator. The electric field variations are in the plane of the antenna, with the magnetic field in a plane at right angles to it, as shown in Figure 8.12.

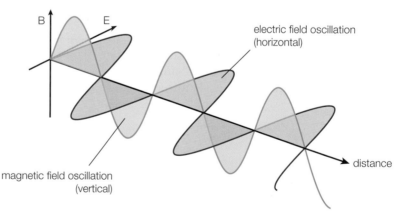

Figure 8.12 ▶
Electromagnetic wave

Electric and magnetic fields will be studied in some detail in Unit 4 of the A2 course, but you should be aware that the fields do not require a material medium and that varying electric fields cannot exist without associated magnetic fields and vice versa. Electromagnetic waves can be detected when they pass over charged particles. Some of their energy is transferred to the particles, which makes them vibrate at the frequency of the wave. Electromagnetic waves can also occur as a result of 'quantum jumps' of electrons in atoms or from 'excited' nuclei. The waves produced in this manner are emitted as **photons**, which are 'bundles' of waves of a few nanoseconds duration that have wavelengths dependent on their energy. The photon nature of light will be investigated in Chapter 10.

The complete range of electromagnetic waves and some of their properties are given in Table 8.1.

Type of wave	Wavelength range/m	Method of production	Properties and applications
γ rays	10^{-16}–10^{-11}	Excited nuclei fall to lower energy states	Highly penetrating rays Used in medicine to destroy tumours, for diagnostic imaging and to sterilise instruments
X-rays	10^{-14}–10^{-10}	Fast electrons decelerate after striking a target	Similar to γ rays but the method of production means that their energy is more controllable Used in medicine for diagnosis and therapy and in industry to detect faults in metals and to study crystal structures
Ultraviolet	10^{-10}–10^{-8}		Stimulates the production of vitamin D in the skin, which results in a tan Makes some materials fluoresce Used in fluorescent lamps and to detect forged banknotes
Visible light	4–7×10^{-7}	Electrons in atoms raised to high energy states by heat or electric fields fall to lower permitted energy levels	Light focused onto the retina of the eye creates a visual image in the brain Can be detected by chemical changes to photographic film and electrical charges on the charge-coupled devices (CCDs) in digital cameras Essential energy source for plants undergoing photosynthesis
Infrared	10^{-7}–10^{-3}		Radiated by warm bodies Used for heating and cooking and in thermal imaging devices
Microwaves	10^{-4}–10^{-1}	High frequency oscillators such as a magnetron Background radiation in space	Energy is transferred to water molecules in food by resonance at microwave frequencies Used in mobile phone and satellite communications
Radio	10^{-3}–10^{5}	Tuned oscillators linked to an aerial	Wide range of frequencies allows many signals to be transmitted Groups of very large radio-telescopes can detect extremely faint sources in space

Table 8.1 ▲

8.3 The wave equation

The speed of a travelling wave depends on the nature of the wave and the medium through which it is passing. However, for all waves there is a relationship between the speed, wavelength and frequency:

$$\text{speed, } v = \frac{\text{distance}}{\text{time}}$$

A wave will travel a distance of one wavelength, λ, in the time it takes to complete one cycle, T:

$$v = \frac{\lambda}{T}$$

In Section 8.1, you saw that the frequency and period of an oscillation were related by the expression:

$$f = \frac{1}{T}$$

It follows that:

$$v = f\lambda$$

that is, wave speed = frequency × wavelength

This is known as the wave equation.

Worked example

1 Calculate the wavelengths of:
   a) the sound from a trumpet playing a note of frequency 256 Hz
   b) ultrasound of frequency 2.2 MHz passing through body tissue.

2 Calculate the frequencies of:

   a) radio waves of wavelength 246 m

   b) microwaves of wavelength 2.8 cm.

Answer

1 a) Speed of sound in air (at 20°C) is 340 m s^{-1}

$$\lambda = \frac{v}{f} = \frac{340\,\text{m s}^{-1}}{256\,\text{Hz}} = 1.33\,\text{m}$$

   b) Speed of sound in soft tissue is 1540 m s^{-1}

$$\lambda = \frac{v}{f} = \frac{1540\,\text{m s}^{-1}}{2.2 \times 10^6\,\text{Hz}} = 0.70\,\text{mm}$$

2 a) Speed of electromagnetic waves (in a vacuum) is 3.0×10^8 m s^{-1}

$$f = \frac{v}{\lambda} = \frac{3.0 \times 10^8\,\text{m s}^{-1}}{246\,\text{m}} = 1.2\,\text{MHz}$$

   b) $f = \dfrac{v}{\lambda} = \dfrac{3.0 \times 10^8\,\text{m s}^{-1}}{0.028\,\text{m}} = 11\,\text{GHz}$

REVIEW QUESTIONS

1 Two oscillations are said to be in antiphase if their phases differ by:

A $\frac{\pi}{2}$ radians

B π radians

C $\frac{3\pi}{2}$ radians

D 2π radians

2 Which of the following does **not** apply to sound waves?

A They always travel at $340\,\mathrm{m\,s^{-1}}$ in air.

B They are longitudinal waves.

C They result from vibrations.

D They transmit energy.

3 Which of the following does **not** apply to X-rays?

A They are produced by fast-moving electrons striking a target.

B They can be detected using a photographic film.

C They have a longer wavelength than visible light.

D They travel through a vacuum with a speed of $3 \times 10^8\,\mathrm{m\,s^{-1}}$.

4 Microwaves of wavelength 12 cm have a frequency of:

A 25 MHz   B 36 MHz

C 2.5 GHz   D 3.6 GHz

5 Define the terms **amplitude**, **frequency** and **period** of an oscillation. Give the SI unit for each.

6 Describe the differences between longitudinal and transverse waves. Give one example of each.

7 The oscilloscope traces in Figure 8.13 are for two oscillations of the same frequency. On the y axis, one division represents a displacement of 1.0 cm; the time base (x axis) is set at 10 ms per division.

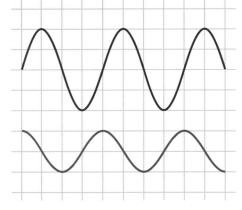

Figure 8.13 ▲

Use the traces to find:

a) the frequency of the oscillations

b) the amplitude of each oscillation

c) the phase difference between the oscillations.

8 Explain the term **wavelength** of a wave. Show that speed, v, frequency, f, and wavelength, λ, are related by the expression $v = f\lambda$.

9 In what ways do electromagnetic waves differ from mechanical waves?

10 a) Calculate:

i) the speed of water surface waves of frequency 50 Hz and wavelength 1.5 cm

ii) the wavelength of electromagnetic radiation of frequency $5.0 \times 10^{15}\,\mathrm{Hz}$

iii) the frequency of microwaves of wavelength 2.8 cm.

b) In what region of the electromagnetic spectrum are the waves in part **ii)**?

9 Transmission and reflection of waves

In Chapter 8, you saw that waves can transfer energy from a source to an observer. In this chapter you will look into the ways different types of wave are affected by the medium through which they travel and what can occur when a wave is incident on an interface between different media.

You will study the properties reflection, refraction and polarisation of waves and how these are used in the food industry and for ultrasound imaging in medicine.

9.1 Transmission

Longitudinal waves

Longitudinal waves progress by the interaction of particles oscillating along the direction of travel of the waves. For sound in air, the successive compressions and rarefactions were shown in Section 8.1 by the candle flame flickering in front of the loudspeaker.

The method of energy transfer can be demonstrated using mechanics trolleys connected together by springs, as shown in Figure 9.1. A pulse applied to the first trolley is transferred from trolley to trolley by the springs and is seen to move along the line of the trolleys. The speed of the pulse can be calculated by dividing the distance between the first and last trolley by the time it takes to travel from start to finish.

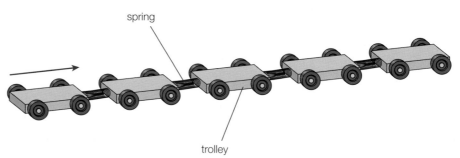

spring

trolley

Figure 9.1 ▲
Transmission of a pulse through trolleys

You will probably already have measured the speed of sound in air by timing the delay between the flash of a gun and the bang over a measured distance. The value is about $340\,\mathrm{m\,s^{-1}}$ at $20\,°C$. For solids such as steel, the value is much bigger, as strong bonds between the particles enable the energy to be transmitted more effectively.

Transverse waves

How does a particle moving sideways enable a wave to progress perpendicularly to the oscillation? This question can be answered using the model of masses connected by springs and the resolution of forces discussed in Chapter 8.

Experiment

Finding the speed of sound in a metal rod

When the steel rod in Figure 9.2 is tapped with the hammer, the circuit is completed and the signal is displayed on the cathode ray oscilloscope (CRO). The pulse travels rapidly along the rod and is reflected back to the hammer. The impulse pushes the hammer away from the rod, which breaks the circuit. The time of contact – the time for the pulse to travel the length of the rod and back again – is found by measuring the length of the signal on the screen and using the time base setting of the CRO. The speed is calculated by dividing the distance travelled – twice the length of the rod – by the time.

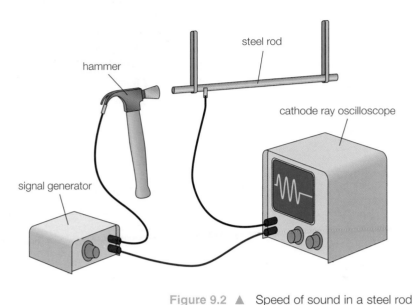

Figure 9.2 ▲ Speed of sound in a steel rod

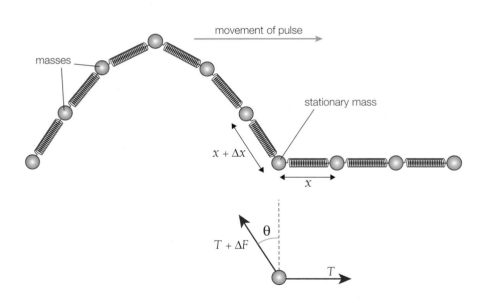

Figure 9.3 ▲

There is an increase in tension between the displaced mass and the stationary mass just ahead of it. Both experience a force that can be represented by two components at right angles. The stationary particle is pulled upwards by the component of the extra tension ($\Delta F \cos \theta$). An analysis using Newton's laws of motion shows that the pulse moves along the row of masses with a speed dependent on the magnitude of the masses and the tension in the springs. This relationship is important when studying the factors that affect the frequency of stringed instruments, which will be investigated in Chapter 10.

Water waves are more complex, because the particles near the surface rotate so that up and down motion of the surface will progress. The speed of surface waves also depends on the elastic properties and mass of the particles, as well as the depth of the liquid.

Experiment

Studying the relationship between the speed of surface waves and the depth of the water

Pour water into a tray to a depth of one centimetre (Figure 9.4). A pulse can
be given to the water by rapidly raising and lowering one end of the tray.
Measure the time taken for the pulse to travel to the end of the tray and back
again. Find the length of the tray and calculate the speed of the wave. Repeat
several times to find an average speed and take further measurements for a
range of depths. Try to deduce a relationship between the speed of the
waves and the depth of the water.

Figure 9.4 ▼ Speed of water waves

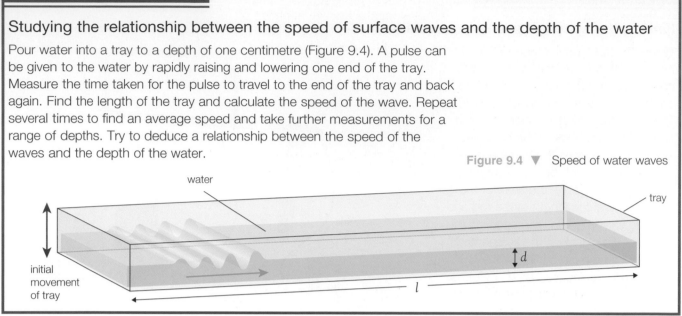

Electromagnetic waves

Unlike mechanical waves, electromagnetic waves do not require a medium for
their transmission. All the regions of the electromagnetic spectrum travel
through a vacuum with a speed of $3 \times 10^8 \, \text{m s}^{-1}$. The radiation interacts with
charged particles and it may be reflected, absorbed or transmitted through
different media. Light, for example, can pass through glass and some plastics.
The speed of the light is reduced by its interaction with electrons and is
dependent on the atomic structure of the medium. The frequency of the wave
affects its speed, with blue light travelling more slowly than red light in media
other than a vacuum.

Some materials reflect or absorb a range of wavelengths but transmit others.
For example, a blue stained-glass window transmits blue light and absorbs
light of all other colours.

The greenhouse effect is due to the upper atmosphere being transparent to
short-wave infrared radiation from the Sun but reflecting back the longer
wavelength rays that are re-radiated from the warm surface of the Earth.

9.2 Reflection

When a wave is incident on an interface between two different media, the
energy may be absorbed, transmitted or reflected. The fraction reflected
depends on the nature of the two media and the angle at which the wave
strikes the interface.

Figure 9.5 shows the paths taken by a ray of light (for example, from a laser)
directed at the surface of a glass block. As the angle is increased, a greater
proportion of the light is reflected.

9.3 Refraction

Figure 9.5 illustrates how a ray of light at an angle to the normal changes
direction when it passes from one medium to another. This effect is due to the
change in wave speed and is known as **refraction**.

The cause of the deviation can be explained using a ripple tank as shown in
Figure 9.6. Waves are generated on the surface of water in a flat-bottomed
glass dish using an oscillating horizontal bar just touching the surface of the

Definition

The law of reflection states that the
angle between the incident ray and a
normal drawn at the point of reflection
is equal to the angle between the
reflected ray and the normal in the
plane of the reflection.

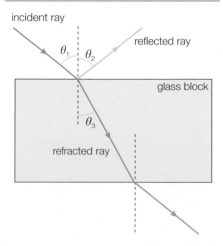

Figure 9.5 ▲

liquid. A series of straight ripples move away from the oscillator. A shadow image of the ripples can be observed using an overhead projector. Each bright line represents the crest of a wave, so all points along the line are in phase. Such a line is called a **wavefront**. The distance between successive bright lines is therefore one wavelength. If a strobe lamp of frequency close to or equal to that of the oscillator is used, the movement of the wavefronts can be slowed down or frozen.

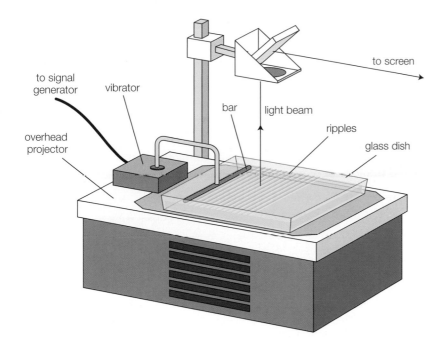

Figure 9.6 ◄
Ripple tank

A piece of clear Perspex® placed in the dish creates a region of shallow water above it. The waves travel more slowly over the shallow water, and as the waves have a fixed frequency the wavelength is reduced ($\lambda = \frac{v}{f}$).

The reduction in wavelength causes the wavefronts to change direction (Figure 9.7). As the wavefronts are perpendicular to the motion, the path of the waves is deviated towards the normal when the speed is reduced and away from the normal when it is increased.

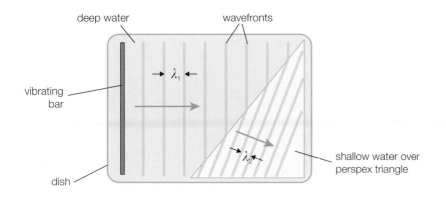

Figure 9.7 ▲   Refraction of wave fronts

For all refracted waves the path is deviated **towards the normal** when the wave is **slowed down** and **away from the normal** when the speed **increases**. In Figure 9.5, the light entering the glass block bends towards the normal on entering the block and away from the normal on leaving. The light travels more slowly in the glass than in the air. The size of the deviation of the wave path depends on the relative speeds in the two media. The ratio of the speeds is called the **refractive index** between the media:

$$\text{refractive index from medium 1 to medium 2} = \frac{\text{speed in medium 1}}{\text{speed in medium 2}}$$

$$_1\mu_2 = \frac{v_1}{v_2}$$

Analysis of the wavefront progression shows that the ratio of the speeds is equal to the ratio of the incident angle and the refracted angle:

$$_1\mu_2 = \frac{\sin\theta_1}{\sin\theta_2}$$

This is known as **Snell's law**.

Worked example

The speed of light is $3.0 \times 10^8\,\text{m s}^{-1}$ in air, $2.3 \times 10^8\,\text{m s}^{-1}$ in water and $2.0 \times 10^8\,\text{m s}^{-1}$ in glass.

1 Calculate the refractive index for light passing from air to water, air to glass and from water to glass.

2 Calculate the angles θ_w and θ_g for light incident at 40° to the normal at the air–water interface in Figure 9.8.

Answer

1 $_a\mu_w = \dfrac{3.0 \times 10^8\,\text{m s}^{-1}}{2.3 \times 10^8\,\text{m s}^{-1}} = 1.3$

$_a\mu_g = \dfrac{3.0 \times 10^8\,\text{m s}^{-1}}{2.0 \times 10^8\,\text{m s}^{-1}} = 1.5$

$_w\mu_g = \dfrac{2.3 \times 10^8\,\text{m s}^{-1}}{2.0 \times 10^8\,\text{m s}^{-1}} = 1.2$

2 $1.3 = \dfrac{\sin 40°}{\sin \theta_w}$

$\theta_w = 30°$

$1.2 = \dfrac{\sin 30°}{30 \sin \theta_g}$

$\theta_g = 24°$

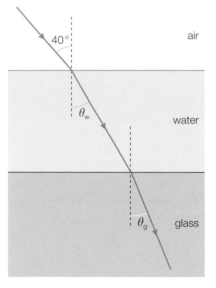

Figure 9.8 ▲

Refraction occurs for all waves. Sound can be deviated as it passes from warm air to cooler air and microwaves can be refracted by wax. It is simpler to demonstrate the refraction of light, however, so the bulk of the applications and effects discussed in this section will relate to the visible region of the electromagnetic spectrum.

In most cases, we observe the effects of light passing across an interface between air and the refracting medium. It is convenient to ignore any reference to the air and state the value as the refractive index of the material (sometimes called the **absolute refractive index**).

The angle of incidence is usually represented by i and the angle of refraction by r, so Snell's law gives the refractive index of a medium by the expression:

$$\mu = \frac{\sin i}{\sin r}$$

For light travelling from a medium of refractive index μ_1 to one of refractive index μ_2 at angles θ_1 and θ_2 (Figure 9.9), a more general expression for Snell's law can be derived using the speeds v_1 and v_2:

$$\mu_1 = \frac{v_a}{v_1} \qquad \mu_2 = \frac{v_a}{v_2}$$

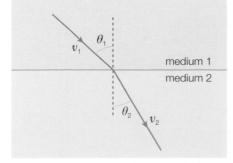

Figure 9.9 ▲

where v_a = speed of light in air

$$\frac{\mu_2}{\mu_1} = \frac{v_1}{v_2} = {_1}\mu_2 = \frac{\sin\theta_1}{\sin\theta_2}$$

$$\mu_1 \sin\theta_1 = \mu_2 \sin\theta_2$$

Experiment

Measuring the refractive index of glass

The glass block in Figure 9.10 is placed on a sheet of plain paper. By tracing the rays through the block, a range of values of i and the corresponding values of r are measured. The gradient of a graph of $\sin i$ against $\sin r$ is the average value of the refractive index of the glass.

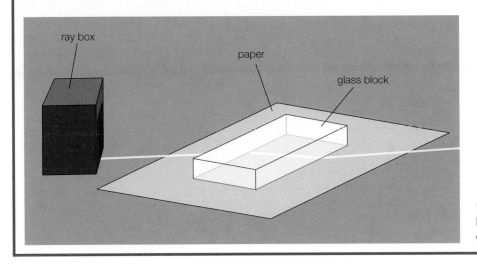

Figure 9.10 ◄
Measuring the refractive index of glass

Total internal reflection

When light travels from glass to air or glass to water, it speeds up and bends away from the normal. The passage of light from a medium with a high refractive index to one with a lower refractive index is often referred to as a 'dense-to-less-dense' or 'dense-to-rare' transition. This relates to optical density, which is not always equivalent to physical density.

If the angle of incidence at the glass–air interface is increased, the angle of refraction will approach 90°. If the angle is increased further, no light can leave the glass and so it is all reflected internally according to the laws of reflection (Figure 9.11).

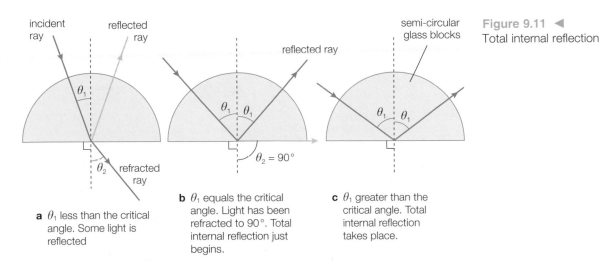

a θ_1 less than the critical angle. Some light is reflected

b θ_1 equals the critical angle. Light has been refracted to 90°. Total internal reflection just begins.

c θ_1 greater than the critical angle. Total internal reflection takes place.

Figure 9.11 ◄
Total internal reflection

93

The angle at which **total internal reflection** just occurs is termed the **critical angle**.

Applying the general form of Snell's law:

$$\mu_1 \sin\theta_1 = \mu_2 \sin\theta_2$$

where μ_1 = refractive index of glass, μ_2 = refractive index of air = 1, θ_1 = critical angle, C, and $\theta_2 = 90°$, leads to:

$$\frac{\mu_1}{\mu_2} = \frac{\sin\theta_2}{\sin\theta_1}$$

Hence, for a glass–air interface:

$$\mu_1 = \frac{1}{\sin C}$$

Worked example

The refractive index is 1.50 for glass and 1.33 for water. Calculate the critical angle for light passing from:

1 glass to air

2 water to air

3 glass to water.

Answer

1 $\sin C = \dfrac{1}{1.50}$ $\qquad$ $C = 42°$

2 $\sin C = \dfrac{1}{1.33}$ $\qquad$ $C = 49°$

3 $\sin C = \dfrac{1.33}{1.50}$ $\qquad$ $C = 62°$

Total internal reflection plays a big part in our lives. Figure 9.12 shows light being reflected back using a prism and the passage of a ray of light through a glass fibre. The prism principle is used for the reflection of light from car headlamps (cat's eyes), and glass fibre is used extensively in communications – a topic that will be studied in detail in Unit 4 of the A2 course.

Figure 9.12 ▶

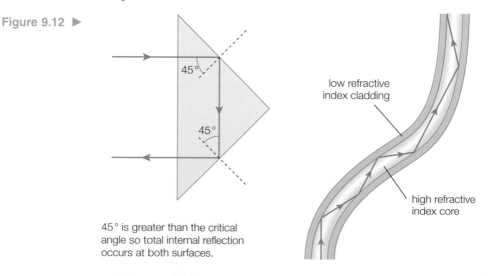

45° is greater than the critical angle so total internal reflection occurs at both surfaces.

low refractive index cladding

high refractive index core

Refractive index of liquids

It is difficult to trace a ray of light through a liquid without the light having to pass through a container. A method involving the measurement of the critical angle at a glass–liquid interface may be used (Figure 9.13).

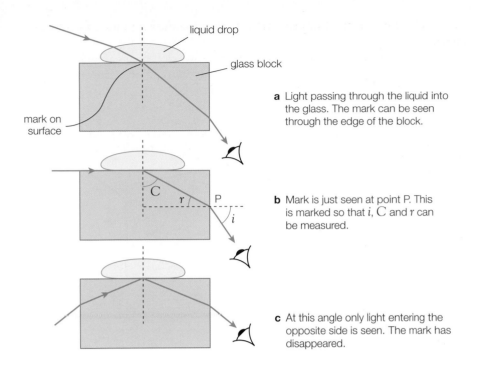

a Light passing through the liquid into the glass. The mark can be seen through the edge of the block.

b Mark is just seen at point P. This is marked so that i, C and r can be measured.

c At this angle only light entering the opposite side is seen. The mark has disappeared.

Figure 9.13 ▲

Using the values of i and r, the refractive index of the glass is found. The refractive index of the liquid is then calculated using C:

$$\mu_g \sin C = \mu_l \sin 90$$

$$\mu_l = \mu_g \sin C$$

This is the principle of a **refractometer** – a device that is used in the food industry to determine the concentrations of sugar solutions, etc, by measuring the refractive index.

9.4 Polarisation

Transverse waves with the particles oscillating in one plane only are said to be plane polarised.

If a rope is fixed at one end, a wave can be passed along it by an up-and-down or side-to-side motion of the hand (Figure 9.14). The waves will lie in a vertical or horizontal plane and are said to be vertically or horizontally **plane polarised**.

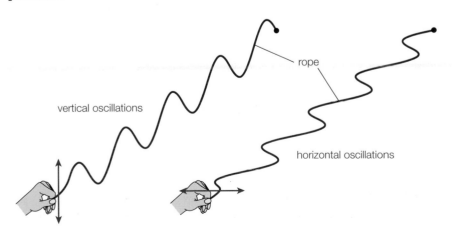

Figure 9.14 ▲

The plane of polarisation can be varied by moving the hand in different directions.

If a rope is passed through a vertical slit in a wooden board (Figure 9.15), horizontal vibrations would be stopped and only vertical ones transmitted. Such an arrangement acts as a **polarising filter**.

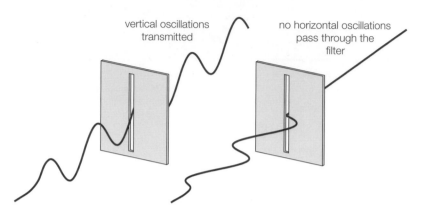

vertical oscillations transmitted

no horizontal oscillations pass through the filter

Figure 9.15 ▲ Polarising filter

Longitudinal waves have particles that have no components of oscillations in the planes perpendicular to the direction and so cannot be polarised. A sound wave, for example, would pass through the slit whatever its orientation.

Radio waves and microwaves transmitted from an aerial have variations in the electric field in the plane of the aerial and in the magnetic field at right angles to it. To pick up the strongest signal, the receiving aerial must be aligned in the same plane as the transmitter (Figure 9.16). When the receiver is rotated, the strength of the received microwaves falls until it reaches zero when it is at right angles to the transmitting aerial. The receiver will only pick up the component of the signal in the plane of its antenna.

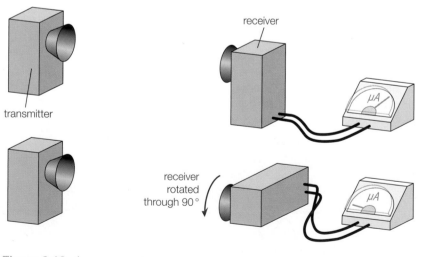

receiver

transmitter

receiver rotated through 90°

Figure 9.16 ▲

Polarised light

Most sources of light are unpolarised. The light emitted from the sun or a lamp consists of variations of electric and magnetic fields in all planes.

Light may be polarised using filters that behave like the slit for the rope and by reflection (Figure 9.17). The filter is usually made of transparent polymers with the molecular chains aligned in one direction. If two filters are held

together and rotated until the transmitting planes are at right angles, a source
of light viewed through them will reduce in intensity and finally disappear.

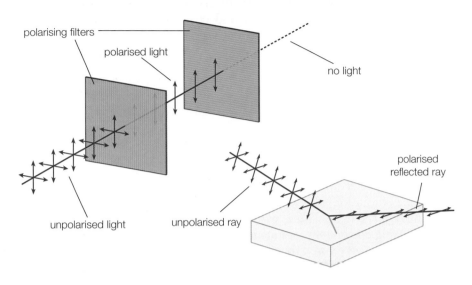

Figure 9.17 ▲

Light falling at an angle onto a transparent material will undergo reflection
and refraction at the surface. At a particular angle of incidence that depends
on the refractive index of the medium, the reflected light will be completely
plane polarised in the plane of the reflecting surface. For glass, this angle is
about 56°. A polarising filter that allows through vertically polarised light will

Experiment

Effect of the concentration of a sugar solution on the plane of polarisation of light

Two polarising filters and a 360° protractor can be used to
measure the effect of the concentration of a sugar solution on the
plane of polarisation of light, but simple polarimeters are widely
available (Figure 9.18).

Distilled water is put into the cell to check that the scale reading is
zero when the filters are crossed and the light-emitting diode (LED)
blacked out. A number of sugar solutions of different
concentrations (20–100 g in 100 ml of distilled water) are prepared.
The cell is filled to a fixed level with the solution. The upper filter is
rotated until the light source disappears. The angle of rotation is
measured. The angle is found for all the known concentrations and
some unknown solutions, including those containing clear honey
and syrup.

A graph of the angle of rotation against the concentration is
plotted and used to find the values for the unknown
concentrations.

Figure 9.18 ▶

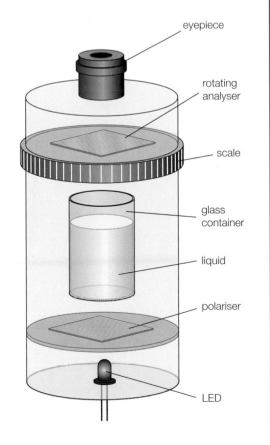

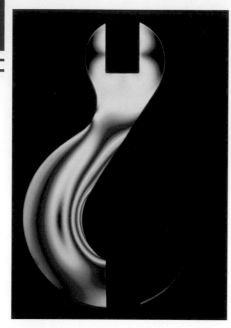

Figure 9.19 ▲
Interference stress patterns

therefore cut out the horizontally polarised reflected light. This is the principle of antiglare sunglasses.

Light can also be polarised by scattering. The sky seems to be blue because the short-wave blue region of the visible spectrum is scattered much more than the red. If you look at the sky through a polarising filter, a difference in intensity can be seen when the filter is rotated. Bee's eyes are able to detect polarised light, and this aids their navigation.

Optical activity

Some complex molecules rotate the plane of polarisation of transmitted light. In sugar solutions, the angle of rotation is dependent on the concentration of the solution.

Stress analysis

Long-chain polymers like Perspex® can also rotate the plane of polarisation. The degree of rotation depends on the strain on the molecules and the wavelength of the light. This is particularly useful in industrial design. Models of load-bearing components are made from Perspex® and are put under stress and viewed through crossed sheets of Polaroid®. The multicoloured interference stress patterns (Figure 9.19) are analysed to detect potential regions of weakness.

9.5 Pulse-echo techniques

One method of finding the speed of sound in air is to bang a drum while standing a measured distance (100 m or more) from the wall of a large building and timing the period between striking the drum and hearing the echo (Figure 9.20).

The speed is calculated by dividing the distance travelled by the sound (to the wall and back) by the time taken. If the speed of the sound in air is known, the same method can be used to find the distance of the observer from the wall:

Figure 9.20 ▶

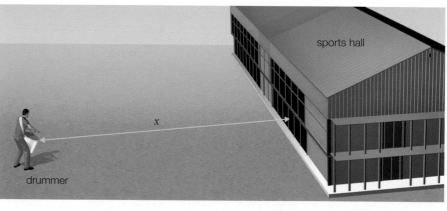

sports hall

x

drummer

$$v = \frac{2x}{t}$$

hence:

$$x = \frac{vt}{2}$$

Sonar and radar are methods developed during the Second World War that are still widely used to gauge the positions of ships and aircraft. They achieve this by sending out pulses of radio and sound waves and noting the times and direction of the reflected pulses. Bats and dolphins are examples of animals that emit and receive high-frequency sounds to navigate or detect food sources.

A more recent development of the pulse-echo technique is ultrasound imaging in medicine. **Ultrasound** describes sound waves of frequency greater than the upper threshold of human hearing (about 20 kHz). In practice, frequencies in the range 1–3 MHz are generally used for medical images.

Amplitude scans (A-scans) are used to determine the depth of boundaries between tissues or bone and tissue (Figure 9.21). Pulses of ultrasound are emitted by a transducer and directed into the body at the region to be investigated. A coupling gel is smeared onto the body at the point of entry so that very little ultrasound is reflected from the skin. Some of the pulse's energy is reflected at the boundaries and received by the transducer.

In Figure 9.21, the reflections from the inner abdomen wall, the front and back of the organ and the spinal column are shown on the screen of a CRO. The time between the reflections and the entry of the pulse can be measured using the time base of the CRO, and the depth of the boundaries calculated using the pulse-echo formula.

Figure 9.21 ◄

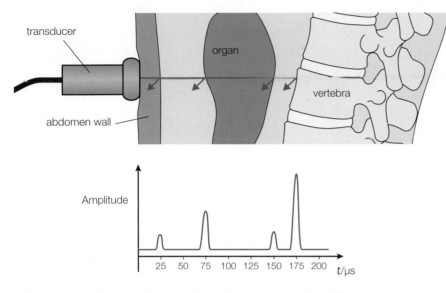

The fraction of sound that is reflected depends on the difference in a property known as the acoustic impedance of the tissue on each side of the interface. The acoustic impedance depends on the density of the medium so a much bigger reflection occurs at a tissue–bone boundary than at a tissue–muscle interface. The amplitude of the reflections received by the transducer will be reduced by attenuation – that is, energy absorbed or scattered within the body. Amplification of the reflected pulses by a factor depending on the distance travelled compensates for the attenuation, and the size of the peaks on the monitor indicates the relative fractions of the pulse reflected at each boundary.

Worked example

Use the timescale of the A-scan in Figure 9.21 to determine the distance of the organ from the inner abdominal wall and the width of the organ. The speed of sound is 1500 m s^{-1} in soft tissue and 1560 m s^{-1} in the organ.

Answer

$$x = \frac{v \, \Delta t}{2}$$

For the tissue:

$$x = \frac{1500 \, \text{m s}^{-1} \times (75 - 25) \times 10^{-6} \, \text{s}}{2} = 3.8 \times 10^{-2} \, \text{m}$$

For the organ:

$$x = \frac{1560 \, \text{m s}^{-1} \times (150 - 75) \times 10^{-6} \, \text{s}}{2} = 5.9 \times 10^{-2} \, \text{m}$$

A-scans can be used to detect a detached retina in the eye. They are also used in echoencephalography to accurately determine the midline of the brain: the midline is the gap between the hemispheres, and any deviation of the midline could indicate the presence of a tumour or haemorrhage on one side of the brain.

B-scans, in addition to detecting the position of the reflecting boundary, give a display where the brightness of the reflection represents the fraction of energy reflected. The familiar image from a foetal scan in Figure 9.22 clearly shows the skeletal structure.

Figure 9.22 ▶
Ultrasound scan of 13-week-old foetus

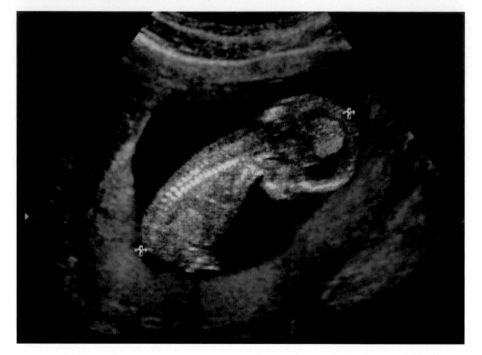

Ultrasound images are generally of lower resolution than X-ray images – that is, they give much less detail. Resolution depends on wavelength, with the shorter wave X-rays giving a clearer, more-detailed image. Medical ultrasound uses wavelengths of a few millimetres, whereas the wavelength of the X-rays used in diagnostic imaging is about 1×10^{-11} m. However, X-ray radiation is ionising and can kill or damage cells in the body. A developing foetus is particularly vulnerable, so ultrasound provides a much safer alternative.

The resolution of ultrasound can be improved by increasing the frequency, but the shorter waves are absorbed more readily and the useful range is reduced. The effective range is about 200 wavelengths, so 3 MHz waves are used for more detailed images of regions close to the skin, while the lower frequencies are used to examine deeper organs.

Doppler effect

You are probably familiar with the change in sound of a racing car as it passes you at high speed: the frequency of the engine noise suddenly falls. Similarly, the frequency of the siren on an emergency vehicle goes up and down as it revolves. This change in frequency due to the relative motion of the source to the observer is known as the **Doppler effect**.

The Doppler effect can be demonstrated using a sphere containing a small oscillator that emits a high-pitched sound attached to a length of string (Figure 9.23). The oscillator is switched on and rotated in a circle on the string. An observer will detect a higher-pitched sound when the sphere approaches and a lower pitch when it moves away.

Figure 9.23 ◄

sound transmitter

listener

Figure 9.24 shows that the wavefronts are compressed into a shorter distance when the source moves towards the observer, which effectively reduces the wavelength and hence increases the observed frequency. When the source is moving away from the observer, the wavelength increases and the frequency is lower. As with most wave representations, many Doppler animations are freely available on the internet.

Figure 9.24 ◄

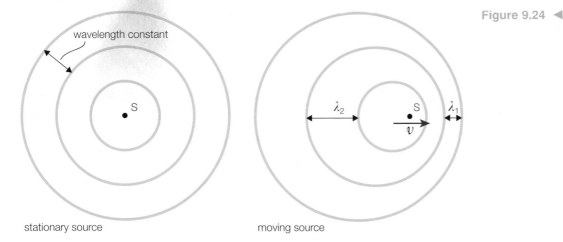

wavelength constant

S

λ_2

S

v

λ_1

stationary source

moving source

The change in frequency of the Doppler effect is dependent on the ratio of the speed of the moving source to the wave speed and the frequency of the stationary source. If the change in frequency, Δf, is measured, the speed of the emitter can be calculated. If the speed of the source, v, is much less than the wave speed, c, the ratio of the change in frequency to the frequency of the source, f, equals the ratio of the speed of the source to the wave speed.

$$\frac{\Delta f}{f} = \frac{v}{c}$$

This relationship will be used in Unit 5 of the A2 course to explain the 'red shift' – an observed reduction in the frequency of known spectral lines that suggests that the universe is expanding.

The 'Doppler shift' occurs whenever there is a relative motion between source and observer: an observer moving towards or away from a stationary source will detect an increase or decrease in frequency. If pulses of short-wave radio waves are directed towards an oncoming vehicle, the vehicle will receive and reflect higher frequency pulses. These are picked up by a receiver with a further increase as they are coming from a moving source. This 'double shift' in frequency can be measured and compared with the transmitted frequency to find the speed of the moving vehicle. This is the principle of police speed traps.

The Doppler effect is commonly used in medicine (Figure 9.25). An ultrasound transducer coupled to the body near an artery emits pulses that are reflected by moving blood cells within the artery. A blockage due to a blood clot (thrombosis) or a constriction caused by the thickening of the arterial wall is detected by a sudden change in the shift frequency. Foetal heartbeats can also be measured safely using a similar method. Pulses reflected off the surface of the beating heart will increase in frequency when the heart wall moves out towards the transducer and will decrease when it moves in, away from the transducer. An oscilloscope trace showing the regular increase and decrease in frequency is used to determine the rate of the foetal heartbeat.

Figure 9.25 ▶
Doppler effect in medicine

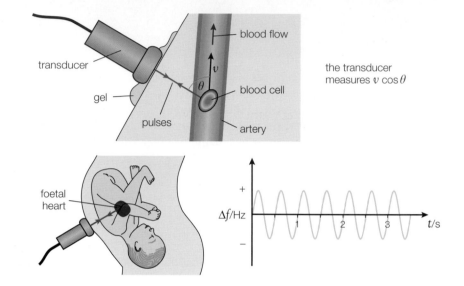

REVIEW QUESTIONS

1 Blue light is deviated more than red light when it enters a glass block because:

  A it has a longer wavelength

  B it has a lower frequency

  C it travels at a greater speed in glass than red light

  D it travels at a lower speed in glass than red light.

2 Light **cannot** be polarised by:

  A passing through a narrow slit

  B passing through sheets of Polaroid®

  C reflecting off a glass surface

  D scattering off dust particles in the atmosphere.

3 Ultrasound is preferred to X-rays for some diagnostic images because:

  A it gives a more detailed image

  B it is a longitudinal wave

  C it is less harmful to the patient

  D it penetrates the body more easily.

4 A firework accelerates upwards and emits a constant high-pitched sound. An observer will hear:

  A a constant higher pitched sound

  B a constant lower pitched sound

  C a continually decreasing pitch

  D a continually increasing pitch.

5 A recorder at the finish line of a 100 metre race sees the flash of the starting pistol and starts her stopwatch. A second timer fails to see the flash and starts his watch on hearing the bang. The winner's time differed by 0.3 s on the two watches. Explain the likely reason for this and use the difference to estimate the speed of sound in air.

6 a) Explain the terms **wavefront** and **wavelength**.

  b) Copy the diagram of light waves travelling from glass to water in Figure 9.26 and continue the passage of the wavefronts into the water.

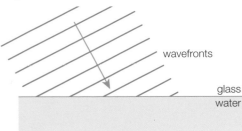

Figure 9.26 ▲

7 a) A ray of light enters one side of a rectangular glass block at an angle of incidence 40°. Calculate the angle of refraction if the refractive index of the glass is 1.55.

  b) The opposite side of the block is immersed in a clear liquid. The ray makes an angle of 28° to the normal when it passes into the liquid. Calculate the refractive index of the liquid.

8 Calculate the critical angle for the interface between the core of an optical fibre with refractive index 1.60 and the cladding with refractive index 1.52.

9 Why can sound waves not be polarised?

10 It has been suggested that cars should be fitted with polarising lenses on their headlamps so that the glare seen by oncoming motorists from the reflections from wet road surfaces will be reduced. Explain how this may be effective.

11 A polarimeter is used to find the angle of rotation of equal lengths of sugar solutions of known concentration. The results are given in Table 9.1.

Concentration/g ml^{-1}	0.20	0.40	0.60	0.80
Rotation angle/°	15	30	46	60

Table 9.1 ▲

Plot a graph of the angle of rotation against concentration and hence deduce the concentration of a similar length of sugar solution that rotates the plane of polarisation through 33°.

12 A trawlerman uses sonar to detect shoals of fish. A strongly reflected pulse is received 1.60 s after it was transmitted. If the speed of sound in water is 1500 m s^{-1}, how far from the boat are the fish?

13 Give one advantage and one disadvantage of using higher frequency ultrasound for diagnostic images in medicine.

14 A space probe sends radio signals back to Earth. Why may the radio receiver at ground control have to be retuned after the launch of the probe?

10 Superposition of waves

When two or more waves of the same type meet at a point, the resultant displacement of the oscillations will be the vector sum of the individual displacements.
In this section you will see how the superposition of waves leads to positions of maximum and minimum amplitude and how coherent sources can create regular regions of high and low intensity called **interference patterns**. You will also study the production and applications of **standing waves**, with particular reference to musical instruments, and **diffraction**, which is the effect of waves being obstructed by objects or apertures.

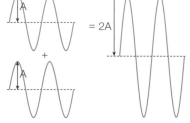

= 2A

constructive superposition

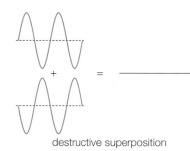

+ =

destructive superposition

Figure 10.1 ▲

Figure 10.1 shows the superposition of two waves in phase and in antiphase (π radians out of phase). The resultant amplitudes are 2 A and zero. When the waves add together to give maximum amplitude, **constructive** superposition occurs; when they combine to produce zero amplitude, **destructive** superposition occurs.

Superposition can only occur for identical wave types. It would not be possible for a sound wave to combine with a light wave, for example. Superposition would also not happen if two transverse waves polarised at right angles were to meet.

Destructive superposition is used in active noise reduction. Sound from a tractor, aeroplane or heavy machinery is picked up using a microphone, electronically processed and transmitted to earphones π radians (180°) out of phase. Destructive superposition takes place, and the noise is effectively cancelled out for the operator.

In most cases, the waves from two or more sources will overlap at a point in a haphazard fashion, so the resulting amplitude variations will not be noticeable. In some situations, however, two sources can produce a pattern of maxima and minima, where the waves combine constructively or destructively at fixed positions relative to the sources. This effect is known as **interference**.

10.1 Interference

Interference patterns

If two identical loudspeakers are placed about one metre apart in a large room with little furniture (for example, a school hall) and both are connected to the same signal generator, an interference pattern will result. An observer walking across the room in front of the speakers will pass loud and quiet

Figure 10.2 ▶
Interference of sound

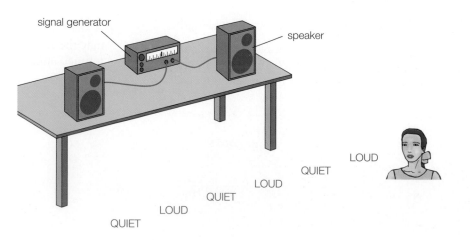

signal generator

speaker

LOUD
QUIET
LOUD
QUIET
LOUD
QUIET

regions regularly spaced as shown in Figure 10.2. Because the speakers are connected to the same signal generator, they will emit sound waves that have identical frequencies and are in phase.

In Figure 10.3, point P_1 is half a wavelength further away from S_2 than S_1. Because the sound from each source is in phase and travels with the same speed, the waves will always be out of phase at P_1, so destructive superposition occurs. Similarly, sound from S_2 will have travelled one whole wavelength further than that arriving at the same time from S_1, so the waves will be in phase at point P_2 and will interfere constructively to give sound of maximum intensity.

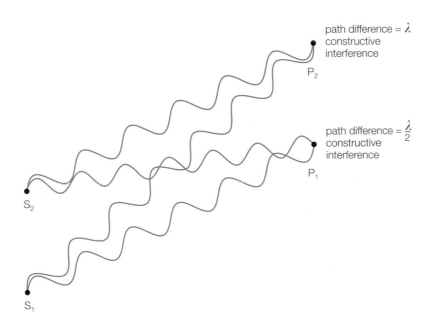

path difference = λ
constructive
interference
P_2

path difference = $\frac{\lambda}{2}$
constructive
interference
P_1

S_2

S_1

Figure 10.3 ◄
Path difference

Definition

Coherent sources have the same frequency and maintain a constant phase relationship.

Tip

Many candidates lose marks by stating that coherent forces are in phase. Any pair of sources could be in phase for an instant so the candidate needs to write 'always in phase'.

The difference in distance from each source to a particular point – for example, $(S_2P_1 - S_1P_1)$ – is called the **path difference**. Positions of maximum amplitude occur when the path difference is zero or a whole number of wavelengths, when the waves are always in phase and constructive superposition takes place. When the path difference is an odd half wavelength, the waves are π radians out of phase and the amplitude will be zero.

Stable interference patterns only occur if:

- the waves are the same type
- the sources are **coherent** (they have the same wavelength and frequency and maintain a constant phase relationship)
- the waves have similar amplitude at the point of superposition.

An interference pattern can be observed using a ripple tank. Circular wavefronts are generated on the surface of the water by two prongs attached to an oscillator. Where a trough from one source meets a trough from the other – or a crest meets a crest – maximum disturbance occurs (Figure 10.4). Calm water indicates the destructive interference where a trough meets a crest.

It is possible to measure the wavelength of radiation using interference patterns. A simple method for measuring the wavelength of microwaves is shown in Figure 10.5a. Aluminium sheets are used to create a pair of slits about 5 cm apart and are placed with their midpoint perpendicular to an aerial and about 10 cm away from a single source. This double slit allows radiation from the same wavefront to pass through both gaps, which effectively creates a pair of coherent sources. The receiving aerial is placed so that it is in line with the transmitter and the midpoint of the slits and is about half a metre to a metre from the slits.

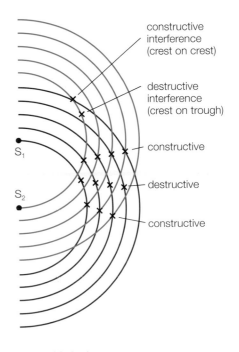

constructive
interference
(crest on crest)

destructive
interference
(crest on trough)

S_1

constructive

S_2

destructive

constructive

Figure 10.4 ▲
Ripple tank interference

Figure 10.5 ▶
a) Apparatus for creating double-slit
interference of microwaves and
b) resultant path differences.

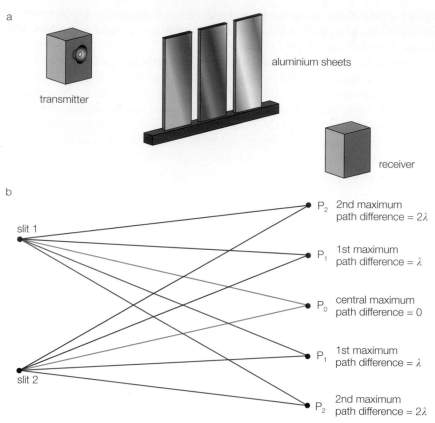

The path difference of the radiation from each slit is zero, so a central maximum signal is received. If the receiver is moved to either side of the slits, a series of alternate maxima and minima will be detected. A sheet of paper is placed on the bench and the position of the aerial at the highest available maximum (usually three or four can be detected) is marked on it. The positions of the slits are then pencilled onto the paper and the distance of each slit from the maximum is measured. The wavelength of the microwaves can be obtained using the relationship:

$$\text{path difference} = S_2 P - S_1 P = n\lambda$$

where n = the order of the maximum from the centre.

Worked example

In an experiment using microwaves, the position of the fourth maximum from the centre was 48 cm from one slit and 60 cm from the other. Calculate the wavelength of the waves.

Answer

Path difference = 12 cm = 4λ

λ = 3 cm

Interference of light

It is quite tricky to observe interference effects with light sources because of the short wavelengths and the difficulty in providing coherent sources.

In 1800, Thomas Young devised a method to produce coherent light sources from wavefronts generated by passing single wavelength (monochromatic) light through a fine slit and then using a double-slit arrangement (like that used in Figure 10.5). The bright and dark lines observed through a microscope eyepiece gave Young the evidence he needed to support his theory of the wave nature of light. The lines are still referred to as Young's fringes.

Nowadays it is much easier to observe interference fringes. Lasers produce intense beams of light that have the same wavelength, are in phase and are polarised in the same plane. A laser beam passed through a pair of parallel lines scored about 1 mm apart on a blackened glass plate will project a fringe pattern onto a wall several metres away (Figure 10.6). Measurements of the slit separation, the distance of the slits from the wall and the fringe width allow the wavelength of the light to be calculated.

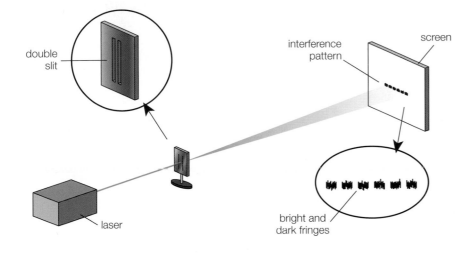

Figure 10.6 ◄
Interference of light

Interferometers

In precision engineering and particularly in the optical industry, surfaces need to be ground with tolerances of better than one thousandth of a millimetre. Interferometers use the patterns created by the recombination of a laser beam that has been split into two separate beams (Figure 10.7). Small changes in the path difference are detected by a shift in the fringe pattern.

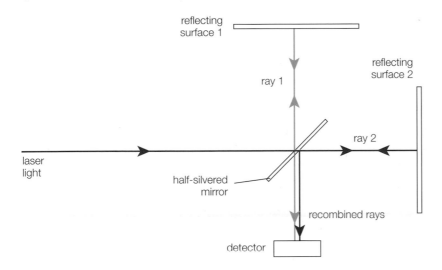

Figure 10.7 ◄
Interferometer

CD players

The information on a CD is in a digital form: the data is stored using binary coding to represent 'bytes' of information in the form of ones and zeros. Electronically these are created by rapidly switching a circuit on and off. Compact discs have a spiral groove with a width of less than 2 μm cut from the edge to the middle of a highly reflective silvered surface. The data is recorded onto the disc as millions of small 'bumps' within the grooves, and the surface of the CD has a clear plastic protective covering.

A laser beam focussed onto the groove is reflected back to a photodiode where the light is converted into an electrical signal (Figure 10.8). Each bump has a height equal to a quarter of the wavelength of the laser (usually about 200 nm, which is a quarter of the wavelength of radiation in the red/infrared region of the electromagnetic spectrum). When the laser illuminates the edge of a bump, the path difference between light reflected from the top of the bump and that from the bottom of the groove is equal to half a wavelength, so the waves interfere destructively and the output of the photodiode is zero (binary 0). When the entire beam is reflected from the upper surface or the gap between the bumps, the intensity of the reflected beam is strong and a high electrical output (binary 1) is generated by the photodiode. The disc rotates rapidly so that the beam follows the groove outwards and the data on the disc is collected as a stream of binary digits.

Figure 10.8 ▶
Principle of CD player.
In practice the beam is much narrower and perpendicular to the surface of the disc.

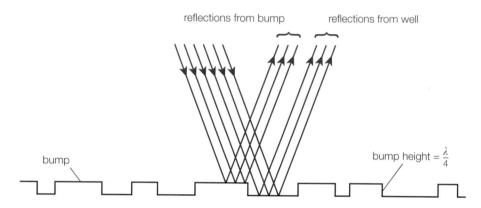

The principle of the CD player can be demonstrated using a standard microwave kit. Most microwave kits have a wavelength of about 3 cm. The 'disc' is made from a wooden board, with 'bumps' of wood 0.75 cm thick glued along its length. The whole length of the board is covered with a strip of aluminium foil, and the board is mounted onto a pair of dynamics trolleys, as shown in Figure 10.9. When the board is pushed across the transmitter/receiver, a sequence of high and low signals is observed.

Figure 10.9 ▶
Microwave model of CD player

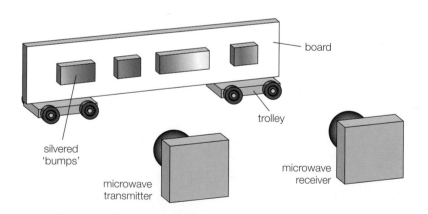

10.2 Standing waves

Standing waves, sometimes called stationary waves, are created by the superposition of two progressive waves of equal frequency and amplitude moving in opposite directions.

   If two speakers, connected to the same signal generator, face each other, a standing wave will exist between them (Figure 10.10). At P, the midpoint between the speakers, the waves, having travelled the same distance at the same speed, will always be in phase and interfere constructively. At A, $\frac{\lambda}{4}$ from P, the distance from speaker 1 has increased by a quarter of a wavelength, while that from speaker 2 has decreased by the same distance. The path difference ($S_1A - S_2A$) is therefore half a wavelength, the waves are in antiphase and destructive interference takes place. Similarly at B, half a wavelength from P, the waves will be back in phase and produce a sound of maximum intensity. A person walking along the line between the speakers will detect a series of equally spaced maxima and minima along the standing wave.

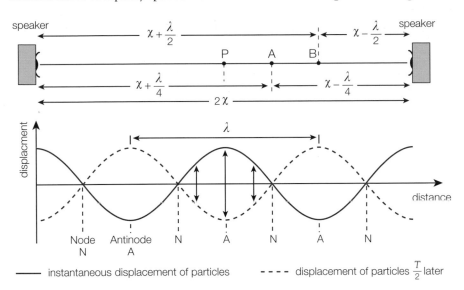

Figure 10.10 ◄
Formation of standing wave

—— instantaneous displacement of particles   - - - - displacement of particles $\frac{T}{2}$ later

The points of zero amplitude within a standing wave are called **nodes**, and the maxima are called **antinodes**. Figure 10.10 shows that the separation of adjacent nodes or antinodes is always half a wavelength.

Experiment

Measuring the wavelength of microwaves

A microwave transmitter is set up in front of an aluminium sheet as in Figure 10.11. Standing waves are formed by the superposition of the reflected wave onto the incident wave. The receiving aerial is moved along the line from the reflector to the transmitter and the distance between ten nodes is measured. This is equal to five wavelengths. The nodes and antinodes are more apparent close to the reflector where the amplitudes of the incident and reflected waves are similar.

Figure 10.11 shows that the same principle can be used to find the speed of sound in the laboratory. The nodes and antinodes are detected using a small microphone placed between a loudspeaker and a reflecting board and connected to a CRO. The wavelength of the sound is measured for a range of frequencies and a graph of λ against $\frac{1}{f}$ is plotted.

The gradient of the graph represents the speed of sound (using $\lambda = \frac{v}{f}$).

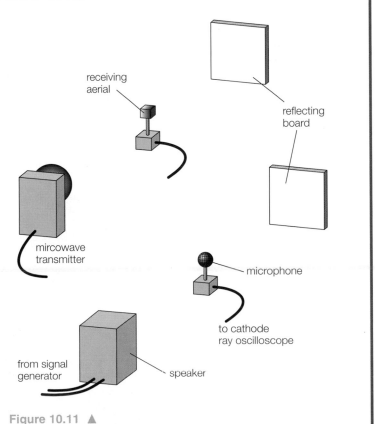

Figure 10.11 ▲
Measuring the wavelength of microwaves and sound using standing waves.

109

Worked example

A standing wave is set up by reflecting a sound of frequency 1200 Hz from a laboratory wall. The distance between four successive nodes is 42 cm. Calculate the speed of sound in the laboratory.

Answer

The distance between nodes is half a wavelength, so there will be 1.5 wavelengths between the first and the fourth node:

$$\frac{3}{2}\lambda = 42\,\text{cm}$$

$$\lambda = 28\,\text{cm}$$
$$v = f\lambda = 1200\,\text{Hz} \times 0.28\,\text{m} = 340\,\text{m s}^{-1}$$

Standing waves differ from travelling waves in the following ways:

- Standing waves store energy, whereas travelling waves transfer energy from one point to another.
- The amplitude of standing waves varies from zero at the nodes to a maximum at the antinodes, but the amplitude of all the oscillations along a progressive wave is constant.
- The oscillations are all in phase between nodes, but the phase varies continuously along a travelling wave.

Standing waves in strings

When a pulse is sent along a rope that is fixed at one end, the reflected pulse is out of phase with the incident pulse. A phase change of π radians (180°) takes place at the point of reflection (Figure 10.12). This means that destructive interference will occur and the fixed position will (not surprisingly) be a node.

Standing waves in a string can be investigated using Melde's experiment (Figure 10.13). A thin length of string is attached to an oscillator, passed over a pulley wheel and kept taut by a weight hanging from its end. The frequency of the oscillator is adjusted until nodes and antinodes are clearly visible. A strobe lamp can be used to 'slow down' the motion so that the standing wave can be studied in more detail.

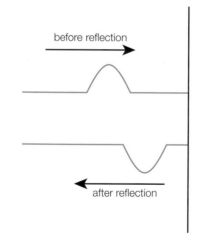

before reflection

after reflection

Figure 10.12 ▲
Reflection of a wave in a string

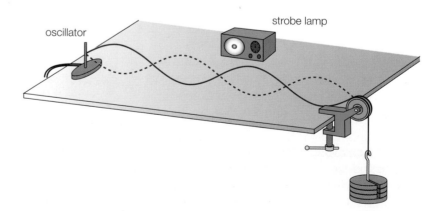

oscillator

strobe lamp

Figure 10.13 ▲ Melde's experiment

Standing waves of shorter or longer wavelength are observed if the frequency of the oscillator is increased or decreased or the weight on the string is altered. The speed of the incident and reflected waves in the string can be calculated using $v = f\lambda$. An investigation into how this depends on the tension or thickness of the string can be performed.

Stringed instruments

Stringed instruments such as guitars, violins and pianos all produce standing waves on strings stretched between two points. When plucked, bowed or struck, the energy in the standing wave is transferred to the air around it and generates a sound. Because the string interacts with only a small region of air, the sound needs to be amplified – either by a resonating sound box or electronically.

The principle of stringed instruments can be demonstrated using a **sonometer** (Figure 10.14). When the string is plucked at its midpoint, the waves reflected from each end will interfere to set up a standing wave in the string. As both ends are fixed, they must be nodes, so the simplest standing wave will have one antinode between two nodes – that is, the length of the string will be half a wavelength.

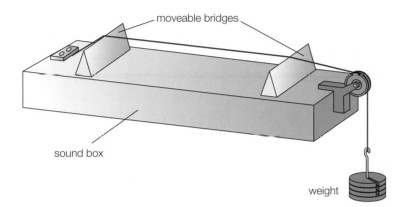

Figure 10.14 ◄
Sonometer

The frequency of the sound emitted from the sonometer can be obtained using a microphone connected to a cathode ray oscilloscope. The relationship between the frequency and the length of the string is found by moving the bridge along the sonometer and noting the frequency of the sound produced by a range of different lengths of wire. A graph of f against $\frac{1}{l}$ is plotted.

A straight line through the origin indicates that the frequency of the sound is inversely proportional to the length of the string.

The sonometer can also be used to investigate the effects of changing the tension in the strings and using wires of different thicknesses and materials. In general, for stringed instruments, the frequency is greater for:

- shorter strings
- tighter strings
- strings that have a lower mass per unit length – that is, thinner strings of the same material.

Overtones and harmonics

You saw earlier that the fixed ends of vibrating strings must be nodes and that the simplest standing wave has a single antinode at the midpoint. The frequency of the note emitted from such a wave is called the **fundamental frequency** of the string. By plucking the string off centre it is possible to create several standing waves on the same string.

Figure 10.15 shows the fundamental mode and two other possible waves. The fundamental vibration has the longest wavelength ($\lambda = 2l$) and the others reduce in sequence. The notes emitted by vibrations other than the fundamental are called **overtones**. Overtones that have whole number multiples of the fundamental frequency are **harmonics**.

The sounds we hear from a guitar, for example, are a complex mixture of harmonics and are noticeably different from the same tune played on a violin. The property that enables us to distinguish different musical instruments is the **quality**, or **timbre**, of the note.

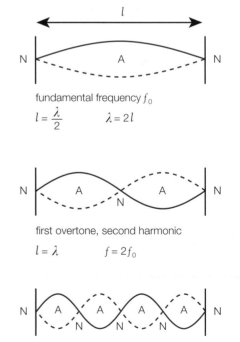

fundamental frequency f_0

$$l = \frac{\lambda}{2} \qquad \lambda = 2l$$

first overtone, second harmonic

$$l = \lambda \qquad f = 2f_0$$

third overtone, fourth harmonic

$$l = 2\lambda \qquad f = 4f_0$$

N = node

A = antinode

Figure 10.15 ▲
Standing waves

Examining the waveforms of stringed instruments

The waveforms of stringed instruments can be examined as in Figure 10.16, in which notes from different instruments are picked up by a microphone connected to a CRO.

A fuller analysis of wave patterns can be achieved using software such as Multimedia Sound. In addition to displaying the sounds played in the room, such software includes a library of pre-recorded notes and a spectrum analyser. The analyser shows the frequency and relative amplitude of the harmonics (Figure 10.17).

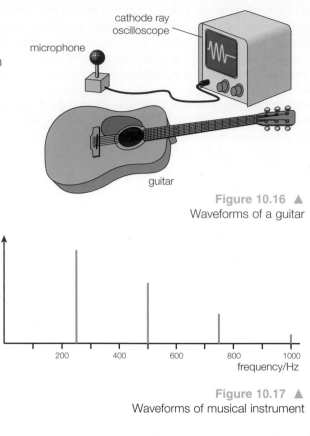

Figure 10.16 ▲
Waveforms of a guitar

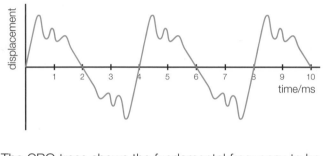

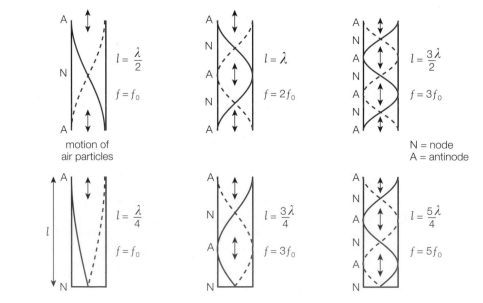

Figure 10.17 ▲
Waveforms of musical instrument

The CRO trace shows the fundamental frequency to be 250 Hz. The analysis reveals that there are overtones of 500 Hz, 750 Hz and 1000 Hz.

Wind instruments

Wind instruments are basically tubes in which standing air waves are formed from vibrations produced in a mouthpiece. Unlike strings, the wave boundaries can be nodes or antinodes.

A small speaker connected to a signal generator can be used to set up standing waves in tubes open at one or both ends. The speaker is clamped just above an open end and the frequency is slowly increased by adjusting the signal generator. When a standing wave is formed in the tube, the air column resonates and an intense booming sound is heard.

Figure 10.18 ▶

Figure 10.18 shows some of the possible wavelengths in a tube open at each end and a tube closed at one end. The waves are traditionally drawn as displacement–distance variations along the tube, but it is important to remember that sound waves are longitudinal and that the vibration of air molecules is along their length.

At the open end, the reflections always create antinodes. At the closed ends, where the particles are unable to oscillate, nodes are formed. The fundamental frequency of the open-ended pipe is therefore twice that of the closed pipe. Figure 10.18 also shows that the first two overtones from the closed tube are the third and fifth harmonics.

Worked example

A small loudspeaker connected to a signal generator emits a sound of frequency 425 Hz. It is fixed above a long glass tube that is filled with water and has a drain at the bottom so that the water can be slowly released from it. When the level has fallen 20 cm from the top of the tube, a standing wave is formed and the air column resonates.

Calculate:

1 the wavelength of the sound

2 the speed of sound in the air column

3 the distance of the water surface from the top of the tube when the next standing wave is formed.

Answer

1 $\dfrac{\lambda}{4} = 20\,\text{cm}$ $\lambda = 80\,\text{cm}$

2 $v = f\lambda = 425\,\text{Hz} \times 0.80\,\text{m} = 340\,\text{m s}^{-1}$

3 Next standing wave formed when $l = \dfrac{3\lambda}{4} = 60\,\text{cm}$ from the top

The notes played on a wind instrument are selected by opening or closing holes along its length and blowing into, or across, a mouthpiece. A recorder is one of the more basic instruments and can be used to demonstrate this principle (Figure 10.19). The air is free to move when the holes are open, so, at resonance, there will be antinodes at the holes. With all stops closed, there will be antinodes at the mouthpiece and the end of the recorder and the wavelength of the lowest fundamental will equal twice the length of the instrument. With some holes open, the fundamental frequency will be that of the standing wave, with antinodes at the mouthpiece and the closest open hole.

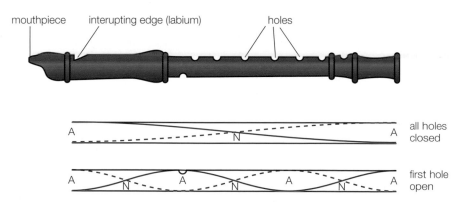

Figure 10.19 ◀
Recorder

Higher harmonics are possible by leaking air into the thumbhole. This effectively divides the air column into two, which allows shorter waves and hence higher pitched notes to be played.

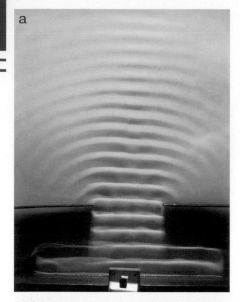

a

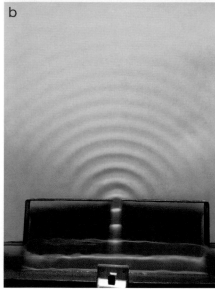

b

Figure 10.20 ▲
Wave diffraction experiment

10.3 Diffraction

When a wave passes through a gap or is partially obstructed by a barrier, the wavefront spreads out into the 'shadow' region (Figure 10.20a). This effect is called **diffraction** and is easily demonstrated using a ripple tank. The dark and light lines represent crests and troughs on the surface of the water, and can therefore be considered as wavefronts.

If the oscillator in a ripple tank is adjusted to a higher frequency, the wavelength shortens and the spreading is reduced. Narrowing the aperture of the gap between two barriers causes even more spreading of the wave (Figure 10.20b). When the width is similar to the wavelength, the wavefronts are almost circular.

Short-wave radio and television signals do not diffract significantly into valleys or around the curvature of the Earth's surface, but military communications radio sends out signals with wavelengths of several kilometres that can be detected by submarines almost halfway around the world.

Close inspection of the water waves on the ripple tank emerging from the narrow aperture in Figure 10.20b reveals regions of constructive and destructive superposition at the edges of the circular wavefront. This effect can be further investigated using a microwave kit (Figure 10.21). The transmitter is placed about 20 cm from a pair of aluminium sheets and directed at a gap of about 5 cm. The receiver is placed facing the aperture and about 50 cm away from the other side of the sheets. The output from the receiver is measured at the midpoint and at 10° intervals as it is moved along a semicircular path.

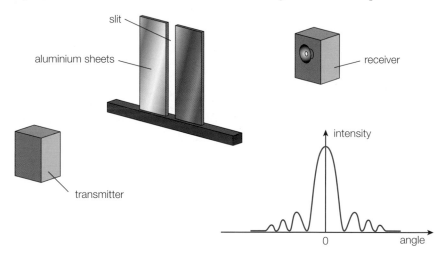

Figure 10.21 ▲

The diffraction pattern shows a central maximum edged by a series of lower intensity maxima and minima as opposed to the regular pattern of interference from a double slit. The central maximum will broaden when the slit width is reduced.

In general, to create a diffraction pattern using a slit, the wavelength should be of the same order of magnitude as the width of the slit.

Diffraction of light

Shadows formed from objects placed in front of a point source of light seem to be sharply defined. There is no noticeable overlap of the light into the shadow region or interference patterns at the edges. The wavelength of light is so short (400–700 nm) that such patterns are difficult to detect with the naked eye. However, many everyday observations are due to diffraction. The star-like pattern produced when light from the Sun, or a bright lamp, is viewed through a fine-mesh material like silk and the multicoloured reflections from the narrow grooves on a compact disc are common examples. Single-slit patterns can be seen between your thumb and forefinger when they are almost touching and between the jaws of vernier callipers.

Diffraction of light can be studied in more detail using a laser. Shadows of small objects and apertures cast onto a screen several metres away clearly show the patterns (Figure 10.22).

Resolution of detail

In Chapter 9 you saw that medical ultrasound images showed much less detail than X-rays. The reason given for this was that the X-rays have a much shorter wavelength than the ultrasound. Although this is valid, it is only part of the explanation. Image resolution is limited by the diffraction that occurs at the aperture of the receiver and so depends not only on the wavelength but also on its diameter.

Figure 10.23 shows the positions of images on the retina from two distant objects. The resolving power of the eye is the smallest angular separation of the objects for which I_1 and I_2 can be separately distinguished. The diffraction patterns formed by the light passing through the pupil of the eye are shown for the images of the objects as their separation decreases. As the images converge, the diffraction patterns begin to overlap on the retina. When the central maximum of one image coincides with the first minimum of the other, they can just be resolved as separate entities, but for smaller angles they merge into a single image. The diameter of the objective lens is much bigger than the pupil, so if the two objects were viewed using a telescope, the central maxima would be narrower and a greater resolution would be possible ($\alpha \approx \dfrac{\lambda}{D}$ where D is the diameter of the aperture).

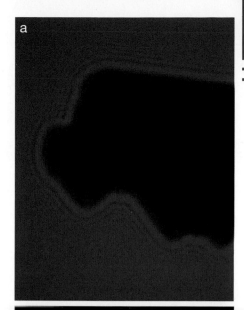

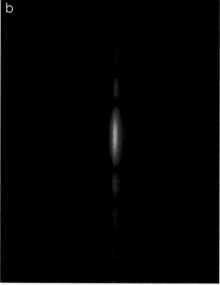

Figure 10.22 ▲
a) Laser light shone on a house key and b) diffraction of laser light shone through a single slit

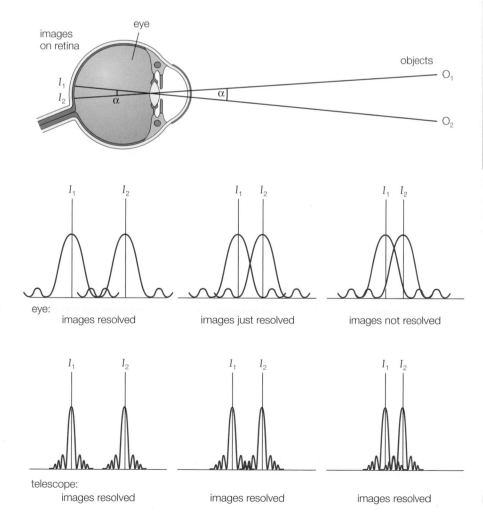

eye:

images resolved   images just resolved   images not resolved

telescope:

images resolved   images resolved   images resolved

Figure 10.23 ◄

Estimating the resolving power of your eye

Two small black dots are marked 1 mm apart on a piece of white paper. The paper is fixed to a wall and you walk away from it until the dots just appear as one. Your distance from the wall, x, is measured. The resolving power,

$$\alpha = \frac{1 \times 10^{-3}\,\text{m}}{x}\ \text{radians}$$

Electron diffraction

In 1912, Max von Laue suggested that if X-rays had a wavelength similar to atomic separations, they should produce diffraction patterns when fired through single crystal materials. He earned a Nobel Prize for his work, and X-ray diffraction techniques have developed so that they now are able to explore the structures of the most complex molecules.

In von Laue's time, electrons had been shown to be particles of mass 9.1×10^{-31} kg carrying a charge of 1.6×10^{-19} C. The idea that particles like electrons could also behave as waves was proposed in 1924 by Louis de Broglie. In Section 16.8, you will use the expression:

$$\lambda = \frac{h}{mv}$$

where h is Planck's constant (6.6×10^{-34} J s).

This equation links the wavelength to the momentum (mv) of the 'particle'. The relationship was subsequently verified using electron diffraction and can be demonstrated using the vacuum tube shown in Figure 10.24. High-velocity electrons are fired from the electron gun at the graphite crystal. A voltage of 1 kV gives the electron a speed of about 2×10^{7} m s^{-1} and hence a de Broglie wavelength of around 4×10^{-11} m. The thin graphite crystal has a regular hexagonal structure with atomic separations similar to this wavelength, so a diffraction pattern of concentric rings is produced on the fluorescent screen. When the voltage across the electron gun is increased, the faster moving electrons have a shorter wavelength so the diffraction is less and the diameter of the rings is reduced.

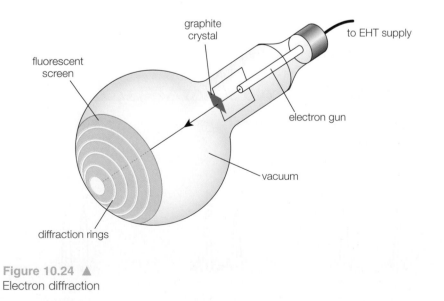

graphite crystal

to EHT supply

fluorescent screen

electron gun

vacuum

diffraction rings

Figure 10.24 ▲
Electron diffraction

REVIEW QUESTIONS

1 Coherent sources always have the same:

A amplitude

B frequency

C intensity

D phase.

2 To form a minimum on an interference pattern, the path difference from two coherent sources that are in phase could be:

A $\dfrac{\lambda}{4}$

B $\dfrac{\lambda}{2}$

C $\dfrac{3\lambda}{4}$

D λ

3 A guitar string has a length of 0.62 m from the bridge to the first fret. The wavelength of the standing wave in the fundamental mode of vibration is:

A 0.31 m

B 0.62 m

C 0.93 m

D 1.24 m

4 The fundamental frequency of the G string on the guitar is 196 Hz. The second harmonic has a frequency of:

A 98 Hz

B 147 Hz

C 294 Hz

D 392 Hz

5 Describe how the principle of superposition is used to reduce the engine noise heard by a fighter pilot. Why are communication signals unaffected by this process?

6 A CD stores data in digital form by encoding it into grooves as a series of bumps of height 160 nm above the base of the groove. Explain how the information is picked up from the CD using a laser and calculate its wavelength.

7 A basic microwave oven with no revolving tray or rotating reflectors can form standing waves by superposition of the transmitted rays and those reflected off the inner surfaces. Food in such an oven will have 'hot spots' and 'cold spots' corresponding to the antinodes and nodes.

A thin slice of cheese cooked in the oven for a short time was found to have small molten regions about six centimetres apart. Estimate the wavelength of the microwaves and show that the microwave frequency is approximately 2.5 GHz.

8 A single filament (festoon) lamp is viewed through a fine slit as shown in Figure 10.25a.

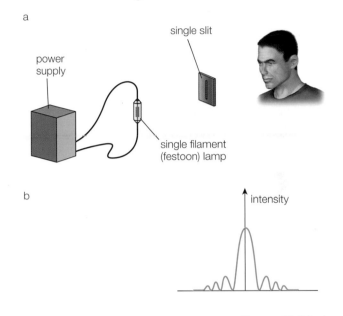

Figure 10.25 ▲

The intensity profile for green light is given in Figure 10.25b.

a) Using the same scales, sketch the diffraction patterns for:

i) a narrower slit

ii) red light with the same slit width

iii) blue light with the same slit width.

b) A second lamp is placed close to and parallel with the first, and the pair are viewed through the slit with a red filter, then a green filter and finally a blue filter. If the green filaments are just distinguishable as separate images, use your sketches to explain the appearance of the red and blue images.

9 A recorder is basically an open pipe in which a standing wave is produced by blowing into the mouthpiece.

a) i) Draw the positions of the nodes and antinodes for a standing wave in an open-ended pipe in the fundamental mode and for the first overtone.

ii) Determine the wavelength in each case.

b) The effective length of a recorder is from the open end at the mouthpiece to the first open hole along its length.

fingers

l

Figure 10.26 ▲

In an experiment to determine the speed of sound in air, the notes played for a range of finger positions were recorded onto a spectrum analyser. The frequencies for each length of tube are given in Table 10.1.

l/m	f/Hz	$\dfrac{1}{l}$ /m⁻¹
0.485	349	
0.435	392	
0.385	440	
0.345	494	
0.325	523	
0.245	698	
0.215	784	

Table 10.1 ◄

i) Complete Table 10.1 and plot a graph of f against $\dfrac{1}{l}$.

ii) Show that $f = \dfrac{v}{2l}$, where v is the speed of sound in air.

iii) Measure the gradient of your graph and hence calculate a value for v.

c) A recorder is said to play sharp (higher pitched) at high air temperatures and flat (lower pitched) at low temperatures. Give a reason for these variations.

11 Charge and current

The Greek philosopher Thales (c. 600 BC) discovered that rubbing the gemstone amber with a cloth caused it to attract small pieces of dry leaf. The Greek word for amber is *elektron*, which is the origin of our words 'electron' and 'electricity'.
In this section you will learn that all matter contains electric **charges** and that if these charges are made to move, an electric **current** is created. The difference between metallic conductors, semiconductors and insulators is discussed in terms of the mobility of charges.

11.1 Electric charge

In simple terms, we can think of all matter as consisting of atoms. These may be considered to be made up of protons and neutrons, which form a nucleus that is surrounded by a 'cloud' of electrons. Protons and electrons have the property of **charge**, which gives rise to electrical forces. Historically, the charge on protons was called **positive** and that on electrons was designated **negative**.

Under normal circumstances we do not observe any effects due to these charges, because most of the time objects have equal numbers of protons and electrons so that the charges cancel out. Indeed, the idea of the **conservation of charge** (that is, equal quantities of positive and negative charge) is a fundamental concept in physics – rather like the conservation of energy. Only when charges **move** in some way does their effect become apparent.

Figure 11.1 confirms that there are two types of charge and that **like charges repel** while **unlike charges attract**. The strips become 'charged' by the transfer of electrons when the strips are rubbed with a duster.

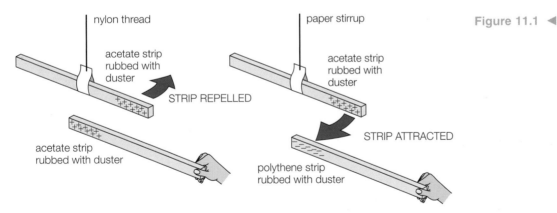

Figure 11.1 ◄

nylon thread

acetate strip rubbed with duster

STRIP REPELLED

acetate strip rubbed with duster

paper stirrup

acetate strip rubbed with duster

STRIP ATTRACTED

polythene strip rubbed with duster

This is possible because the electrons, which are on the outside, can be detached fairly easily from their atoms. The process is as shown in Figure 11.2.

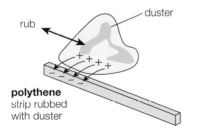
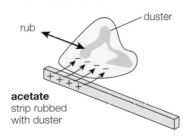

Figure 11.2 ◄

rub — duster

polythene
strip rubbed
with duster

Electrons move from the duster to
the strip, making the strip **negative**
(and leaving the duster positive).

rub — duster

acetate
strip rubbed
with duster

Electrons move from the strip to
the duster, making the strip **positive**
(and leaving the duster negative).

11.2 Electric current

Electric current is defined as the **rate of flow of charge** – that is, the quantity of charge flowing past a given point in one second. Mathematically, we can express this as:

$$I = \frac{\Delta Q}{\Delta t}$$

where I is the current and ΔQ is the amount of charge flowing in time Δt.

Current is measured in **amperes** (symbol **A**). The ampere is the base unit of electric current and is defined in terms of the force between two parallel wires each carrying a current of 1 A. In laboratory terms, an ampere is quite a large current, so we often use mA (milliampere = 10^{-3} A) and μA (microampere = 10^{-6} A).

We can arrange $I = \dfrac{\Delta Q}{\Delta t}$

to give:

$$\Delta Q = I \, \Delta t$$

This tells us that the amount of charge flowing in a certain time is given by multiplying the current by the time. If a current of 1 A flows for 1 s, the quantity of charge flowing is said to be 1 **coulomb** (symbol **C**). The coulomb is a relatively large amount of charge, so we often use μC or even nC ('nano' = 10^{-9}); indeed, the charge on a single electron is only 1.6×10^{-19} C!

> **Tip**
>
> Remember to convert mA and μA to A when working out numerical problems.

> **Tip**
>
> When doing calculations, you must always remember to put the current in amperes, the charge in coulombs and the time in seconds.

Worked example

1 How much charge flows through the filament of an electric lamp in 1 hour when the current in it is 250 mA?

2 The electron charge is 1.6×10^{-19} C. How many electrons flow through the filament during this time?

Answer

1 $\Delta Q = I \, \Delta t$
$= 250 \times 10^{-3}\,\text{A} \times (60 \times 60)\,\text{s}$
$= 900\,\text{C}$

2 Number of electrons $= \dfrac{\text{total charge}}{\text{charge on an electron}}$

$= \dfrac{900\,\text{C}}{1.6 \times 10^{-19}\,\text{C}}$

$= 5.6 \times 10^{21}$ electrons

Experiment

Set up the series circuit shown in Figure 11.3.

You should find that the current is approximately 10 mA, depending on the exact values of the cell and the resistors. You should always remember that the values of resistors are only **nominal values** and that there is a manufacturing tolerance of probably 2% or as much as 5% on the value stated.

Now move the ammeter to point X and then point Y in the circuit. You will observe that (within experimental error) the current is the same wherever the ammeter is in the circuit.

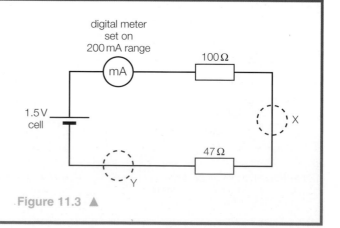

Figure 11.3 ▲

Experiment

Set up the parallel circuit shown in Figure 11.4. Adjust the variable power supply so that the current I at point Z is approximately 5 mA. Without altering the power supply, measure the current, in turn, at points X and Y. Now repeat this procedure by adjusting the power supply to give four more values of I_Z, tabulating I_Z, I_X, I_Y and $I_X + I_Y$.

You should find that, allowing for experimental error, $I_Z = I_X + I_Y$ in all cases. This shows that the current flowing out of a junction is always equal to the current flowing into the junction. This is another example of the conservation of charge – the rate of charge flowing out of a junction must always equal the rate at which it enters the junction because charge cannot be lost or gained.

If you plot a graph of I_Y against I_X, you should find that the gradient is equal to the **inverse** of the ratio of the resistor values (Figure 11.5).

A typical set of results from such an experiment is recorded in Table 11.1.

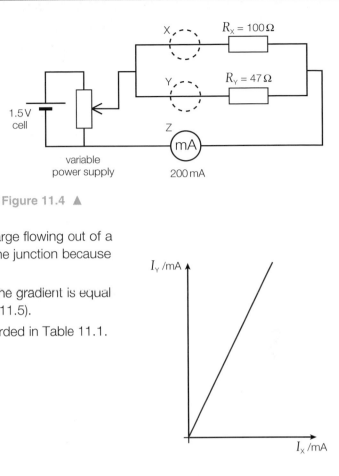

Figure 11.4 ▲

I_Z/mA	I_X/mA	I_Y/mA	$I_X + I_Y$/mA
5.0	1.6	3.4	
15.0	4.9	10.3	
25.0	8.1	16.8	
35.0	11.2	23.7	
45.0	14.5	30.6	

Table 11.1 ◄

Figure 11.5 ▲

Complete Table 11.1 by adding values for $I_X + I_Y$.

Plot a graph of I_Y against I_X and compare the gradient with the ratio of R_X to R_Y.

11.3 Current in series and parallel circuits

The current in a component can be measured by connecting an **ammeter** in **series** with the component. Ammeters have a low resistance so that they do not affect the current that they are measuring.

In a series circuit, the current is the same in each component. This is because the rate at which electrons leave any component must be the same as the rate at which they enter the component – if this were not the case, electrons would be lost from the circuit, which would contravene the principle of conservation of charge (see Section 11.1).

11.4 Drift velocity

In order for a current to flow in a material, suitable **charge carriers** must be present within the material – for example, loosely bound electrons in metals or ions in electrolytes and gases.

The idea of a charge carrier is demonstrated in the experiment in Figure 11.6. When the high voltage is switched on, the metallised sphere swings back and forth between the metal plates, transferring charge as it does so. Figure 11.6b shows that positive charge is carried from left to right and vice versa. This is equivalent to a current flowing clockwise around the circuit, which is recorded by the nanoammeter. In this experiment, the sphere can be considered to be the 'charge carrier'.

> **Tip**
>
> When setting up a series circuit, the ammeter can be placed at any position in the circuit.

> **Tip**
>
> Remember that charge (and therefore current) must always be conserved at a junction.

Figure 11.6 ▶

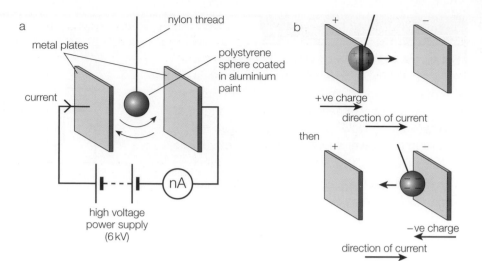

It is conventional to take the direction of an electric current to be the direction in which a positive charge would move. If the charge carriers are negative – for example, the electrons in the connecting wires – the charge carriers actually move in the opposite direction to the current.

Worked example

In the experiment in Figure 11.6, the average current is 30 nA when the sphere takes 16 s to travel across the plates and back a total of 40 times. How much charge is carried on the sphere each time?

Answer

Total charge $\Delta Q = I \, \Delta t = 30\,\text{nA} \times 16\,\text{s} = 480\,\text{nC}$

In this time, the sphere has travelled across and back 40 times – that is, a total of 80 transits. The charge carried on the sphere each time is therefore

$$\frac{480\,\text{nC}}{80} = 6\,\text{nC}$$

(Note that the current has been left in nA, which means that the charge will be given directly in nC. The use of 'quantity algebra' – that is, including the units with each numerical value – will help ensure that errors are not made when doing a calculation in this way.)

In a metallic conductor, the charge carriers are loosely bound outer electrons – so-called 'free' or 'delocalised' electrons. On average, there is approximately one 'free' electron per atom. These 'free' electrons move **randomly** to and fro within the crystal lattice of the metal at speeds approaching one thousandth the speed of light and are also mainly responsible for metals being good conductors of heat. When a potential difference (see Section 12.1) is applied to a circuit, an electric field is created and exerts a force on the 'free' electrons, which causes them to 'drift' in the direction of the force. In accordance with Newton's second law, the electrons would continuously accelerate if it were not for the fact that they collide with the atoms in the lattice (which are, in effect, positive ions, because the 'free' electrons are detached from the atoms, leaving the atoms with a positive charge). These collisions cause an equal and opposite force to be exerted on the electrons, which, by Newton's first law, continue with a constant '**drift velocity**', giving rise to a constant current.

For a conductor, this current is given by:

$$I = nAvq$$

where I = current, A = area of cross-section of conductor, n = number of charge carriers per cubic metre, q = charge on each charge carrier and v = drift velocity of charge carriers.

This may be deduced by considering that the 'volume' of charge carriers passing any point in 1 second is vA (Figure 11.7). The number of charges passing in 1 s is therefore nvA, and if each carries a charge q, the charge passing in 1 s is $nvAq$. By definition (see Section 11.2), this is the current I.

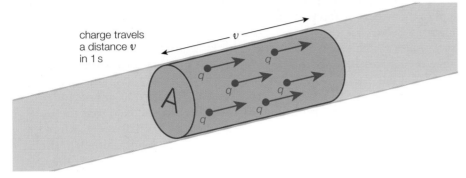

charge travels a distance v in 1 s

Figure 11.7 ▲

Worked example

Consider the circuit in Figure 11.8.
The diameter of the tungsten filament is 0.025 mm and the diameter of the copper connecting leads is 0.72 mm.

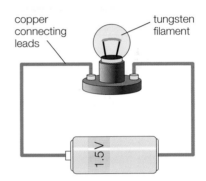

copper connecting leads

tungsten filament

1.5 V

Figure 11.8 ▲

1 Calculate the drift velocity of the electrons in:

  a) the filament

  b) the connecting leads

  when the circuit current is 160 mA.

  You may take the number of charge carriers per cubic metre to be 4.0×10^{28} m^{-3} for tungsten and 8.0×10^{28} m^{-3} for copper.

Answer

1 **a)** For the filament: $I = nAvq$
    Rearranging:

$$v = \frac{I}{nAq}$$

$$= \frac{160 \times 10^{-3}\,\text{A}}{4.0 \times 10^{28}\,\text{m}^{-3} \times \pi \times (0.5 \times 0.025 \times 10^{-3}\,\text{m})^2 \times 1.6 \times 10^{-19}\,\text{C}}$$

$$= 0.051\,\text{m s}^{-1}$$
$$= 51\,\text{mm s}^{-1}$$

  b) For the leads:

$$v = \frac{I}{nAq}$$

$$= \frac{160 \times 10^{-3}\,\text{A}}{8.0 \times 10^{28}\,\text{m}^{-3} \times \pi \times (0.5 \times 0.72 \times 10^{-3}\,\text{m})^2 \times 1.6 \times 10^{-19}\,\text{C}}$$

$$= 3.1 \times 10^{-5}\,\text{m s}^{-1}$$
$$= 0.031\,\text{mm s}^{-1}$$

This worked example shows two rather surprising but very important facts. Firstly, it shows how slow the drift velocity is – particularly in the copper connecting leads. In fact, an electron would probably not get around the circuit before the cell ran out! Secondly, the drift velocity is very much **faster** in the very thin tungsten filament. This is analogous to the flow of water in a pipe speeding up when it meets a constriction – just think of water squirting out at high speed when you put your finger over the end of a tap (this is not recommended as a practical exercise in the laboratory!). As we saw in Section 11.3, the current in a series circuit is the same everywhere, and in order for this to be the case, the electrons in the filament have to speed up, as there are far fewer of them.

Continuing the water analogy, a high pressure difference is set up across the constriction; in electrical terms, a large **potential difference** is developed across the filament (virtually all of the 1.5 V or so provided by the cell), which applies the necessary force to speed up the electrons. This is dealt with more fully in Section 12.1.

Consider switching on an electric light. How is it that the light seems to come on instantly and there is a current in the order of 0.25 A when the drift velocity of the electrons in the wires is only a fraction of a millimetre per second? The answer to the first question lies in the fact that although the electrons themselves are travelling so slowly, the electric field that causes them to move travels at nearly the speed of light ($\approx 3 \times 10^8\,\text{m s}^{-1}$). All of the electrons therefore start to move almost instantly. Secondly, although the individual electrons are moving along the wire very slowly, there is simply an enormous number of them (remember, $n \approx 10^{29}\,\text{m}^{-3}$) and therefore the charge flowing per second equates to a significant current.

Experiment

Set up the arrangement shown in Figure 11.9. Soak the filter paper with 1 M ammonium hydroxide solution and carefully place a small crystal of copper sulphate and a small crystal of potassium permanganate near the centre of the slide. Switch on the power supply.

Observe the movement of the blue copper ions and the purple permanganate ions and estimate their drift speeds. You will find that the drift speed is very slow and the experiment needs to be left to run for an hour or so. What can you deduce about the signs of the respective charges on the copper and permanganate ions?

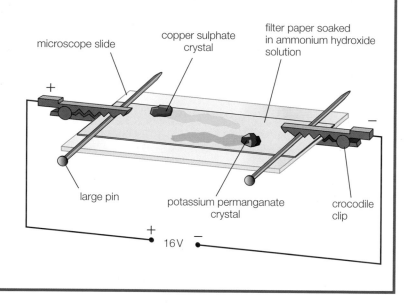

Figure 11.9 ▶

11.5 Metals, semiconductors and insulators

Theories of conduction in different types of material are complex and go well beyond the scope of the A-level course. However, we can give a simple explanation of the electrical conductivity of metals, semiconductors and insulators in terms of the by now familiar equation $I = nAvq$.

The charge q is usually the electron charge, e (or possibly $2e$ for a doubly charged ion), and A is not a property of the material itself. This means that the value of n really determines the ability of a given material to conduct an electric current.

Metals typically have a value of n in the order of $10^{28}\,\text{m}^{-3}$–$10^{29}\,\text{m}^{-3}$, while insulators such as glass and polystyrene have virtually no charge carriers at room temperature. In between, semiconductors such as germanium ($n \approx 10^{19}\,\text{m}^{-3}$) and silicon ($n \approx 10^{17}\,\text{m}^{-3}$) are able to conduct – but not as well as metals. The ability of semiconductors to conduct is greatly enhanced by increasing their temperature or adding 'impurity atoms'. This is discussed more fully in Section 14.5.

REVIEW QUESTIONS

1 The beam current in a television tube is 32 mA. The electrons travel at a speed of $4.2 \times 10^7 \, m \, s^{-1}$ through a distance of 21 cm.

a) The number of electrons striking the screen in 1 hour is:

   A 2.0×10^{17} B 7.2×10^{20}

   C 1.2×10^{22} D 7.2×10^{23}.

b) The number of electrons in the beam at any instant is:

   A 1.0×10^9 B 1.0×10^{11}

   C 1.0×10^{12} D 1.0×10^{14}.

2 A typical flash of lightning lasts for about 0.5 ms, during which time it discharges approximately 8 C of charge.

a) What is the average current in such a strike?

b) The current is caused by positive ions having a charge of $1.6 \times 10^{-19} \, C$ travelling from the Earth to the thundercloud. How many such ions are there in the strike?

3 a) State what physical quantity is represented by each term in the equation:

$$I = nAvq$$

b) A semiconducting strip 6.0 mm wide and 0.50 mm thick carries a current of 10 mA, as shown in Figure 11.10.

6 mm   0.5 mm

10 mA

Figure 11.10 ▲

If the value of n for the semiconducting material is $7.0 \times 10^{22} \, m^{-3}$, show that the drift speed of the charge carriers, which carry a charge of $1.6 \times 10^{-19} \, C$, is about $0.3 \, m \, s^{-1}$.

c) The drift speed for electrons in a copper strip of the same dimensions and carrying the same current would be about $10^{-7} \, m \, s^{-1}$. Use the equation to explain why this value is very different from that of the semiconductor.

4 a) The current in the tungsten filament of a torch bulb is 25 mA.

  i) How much charge will conduct through the filament in 8.0 minutes?

  ii) How many electrons will flow through the filament during this time?

b) The bulb is connected to the cell in the torch by two copper leads. Discuss the factors that would cause a difference between the drift speed of the electrons in the copper leads and that of the electrons in the filament.

12 Potential difference, electromotive force and power

In Section 11 we saw that a moving electric charge gives rise to an electric current. We will now look at how we can make charges move.

You will learn that a cell produces an **electromotive force**. this creates a **potential difference** (or 'voltage') in a circuit, which exerts a force on the charges, causing them to move round the circuit. The potential difference does **work** on the charges, thus developing **power** in the circuit.

12.1 Potential difference

In order to understand how electrons can be made to flow around a circuit and thereby create an electric current, comparison with the flow of water in a pipe is often helpful.

Such an analogy is shown in Figure 12.1.

Figure 12.1 ▶

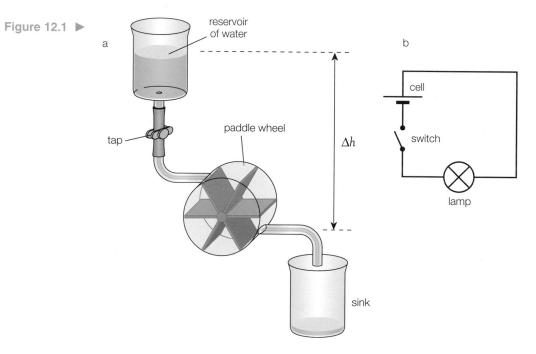

In Figure 12.1a, there is a **pressure difference** between the water level in the reservoir and that in the horizontal pipe. When the tap is opened, this pressure difference causes the water to flow through the pipe and turn the paddle wheel. The **gravitational potential energy** stored in the reservoir is converted into **kinetic energy** of rotation of the paddle wheel.

In Figure 12.1b, a chemical reaction in the cell creates an **electrical potential difference** (symbol V), which exerts a force on the electrons in the wires. When the switch is closed, this force causes the electrons to move around the circuit and through the lamp. Here the **chemical potential energy** stored in the cell is converted into vibrational **kinetic energy** of the atoms in the filament of the lamp, thereby causing its temperature to rise.

In an electric circuit, electrical energy is converted into other forms of energy – for example in the circuit of Figure 12.1b, the lamp converts electrical energy into thermal energy and light. Conversely, an electric motor converts electrical energy into mechanical energy. In our analogy, if the paddle wheel in Figure 12.1a were to be connected to an electric motor, the water could be pumped up again from the sink to the reservoir.

These energy conversions in electrical circuits form the basis of the definition of **potential difference** (often abbreviated to p.d.). The potential difference between two points in a circuit is the amount of electrical energy converted into other forms of energy (thermal, light, mechanical, etc) when unit charge passes from one point to the other. This can be expressed as:

$$\text{potential difference} = \frac{\text{electrical energy converted into other forms}}{\text{charge passing}}$$

or, as energy is the result of doing work, W:

$$\text{potential difference} = \frac{\text{work done}}{\text{charge passing}}$$

From $V = \dfrac{W}{Q}$ it follows that the unit of p.d. is joules per coulomb ($J\,C^{-1}$).

This unit is called the **volt**. Thus, a p.d. of 1 V exists between two points in a circuit if 1 J of energy is converted when 1 C of charge passes between the points. Because potential difference is measured in volts, potential difference is often called the **voltage** between two points in a circuit.

Worked example

The element of an electric kettle that takes a current of 12.5 A produces 540 kJ of thermal energy in three minutes.

1 How much charge passes through the element in these three minutes?

2 What is the potential difference across the ends of the element?

Answer

1 $\Delta Q = I\,\Delta t = 12.5\,\text{A} \times (3.0 \times 60)\,\text{s} = 2250\,\text{C}$

2 $V = \dfrac{W}{Q} = \dfrac{540 \times 10^3\,\text{J}}{2250\,\text{C}} = 240\,\text{V}$

Worked example

A 12 V pump for a fountain in a garden pond can pump water up to a height of 0.80 m at a rate of 4.8 litres per minute.

1 How much work does the pump do per minute when raising the water to a height of 0.80 m? (You may assume that one litre of water has a mass of 1 kg.)

2 If the pump is 75% efficient, how much charge passes through the pump motor in one minute?

3 What current does the motor take when operating under these conditions?

Answer

1 $W = m\,g\,\Delta h = 4.8\,\text{kg} \times 9.8\,\text{m s}^{-2} \times 0.80\,\text{m} = 37.6\,\text{J} = 38\,\text{J}$

2 $V = \dfrac{W}{Q}$

So:

$Q = \dfrac{W}{V} = \dfrac{37.6\,\text{J}}{12\,\text{V}} = 3.1(3)\,\text{C}$ (if 100% efficient)

As the pump is only 75% efficient:

$Q = \dfrac{100}{75} \times 3.1(3)\,\text{C} = 4.2\,\text{C}$

3 $I = \dfrac{Q}{t} = \dfrac{4.2\,\text{C}}{60\,\text{s}} = 70\,\text{mA}$

12.2 Using a voltmeter

The potential difference between two points in a circuit is measured by connecting a **voltmeter** between the points. We talk about connecting a voltmeter **across**, or in **parallel** with, a component to measure the p.d. between its ends.

In the circuit shown in Figure 12.2, the voltmeter is measuring the p.d. across the lamp. To find the p.d. across the resistor, the voltmeter would have to be connected between A and B, and to measure the p.d. across the cell, it would have to be connected between A and C.

> **Tip**
>
> When setting up a circuit, always set up the series part of the circuit first and check that it works. Then connect the voltmeter in the required position.

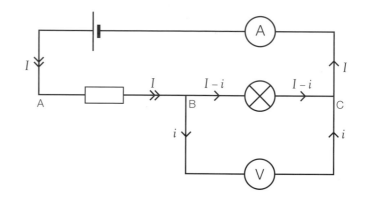

Figure 12.2 ▲

A voltmeter must take some current in order to operate. In the circuit in Figure 12.2, the ammeter records the circuit current, I, but the current through the lamp is only $I - i$, where i is the current taken by the voltmeter. In order to keep i as small as possible, voltmeters should have a very **high resistance**. Typically, a 20 V digital voltmeter might have a resistance of 10 MΩ; analogue meters need more current for their operation and are likely to have resistances in the order of kΩ.

Experiment

Set up the circuit shown in Figure 12.3.

Record the potential difference V across the four-cell power supply in Table 12.1.

V/V	V_1/V	V_2/V	V_3/V	$(V_1 + V_2 + V_3)$/V

Table 12.1 ▲

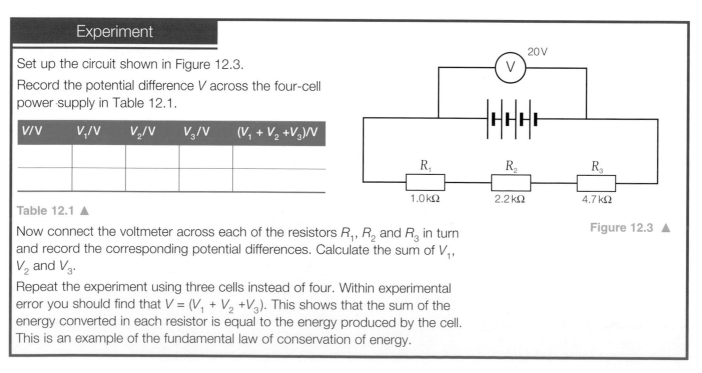

Figure 12.3 ▲

Now connect the voltmeter across each of the resistors R_1, R_2 and R_3 in turn and record the corresponding potential differences. Calculate the sum of V_1, V_2 and V_3.

Repeat the experiment using three cells instead of four. Within experimental error you should find that $V = (V_1 + V_2 + V_3)$. This shows that the sum of the energy converted in each resistor is equal to the energy produced by the cell. This is an example of the fundamental law of conservation of energy.

12.3 Electromotive force

A cell or generator (dynamo or alternator) does work on charges just as a pump does work on water (see the worked example on page 127). Chemical energy is converted into electrical energy in a cell, while mechanical energy (for example, from the engine of a car) is converted into electrical energy in an alternator. The cell or alternator is said to produce an **electromotive force**. This is rather misleading as it is not a force at all but a form of energy transfer! Electromotive force is usually abbreviated to e.m.f. and is given the symbol ε.

The e.m.f. of an electrical source is defined as the energy (chemical, mechanical, etc) converted into electrical energy when unit charge passes through the source. This can be expressed as:

$$\text{e.m.f.} = \frac{\text{energy converted into electrical energy}}{\text{charge passing}}$$

$$\varepsilon = \frac{W}{Q} \text{ or } W = \varepsilon Q$$

As $\dfrac{W}{Q}$ has units of J C^{-1},

the units of e.m.f. are the same as those of potential difference – that is, volts, V.

12.4 Power

In Section 5, we defined power as the **rate of doing work**:

$$P = \frac{\Delta W}{\Delta t}$$

This was based on doing work mechanically. What is the situation if the work is done electrically? From the definition of potential difference:

$$W = QV$$

If both sides of the equation are divided by time, t:

$$\frac{W}{t} = \frac{Q}{t} \times V$$

But $\dfrac{W}{t}$ = power, P

and $\dfrac{Q}{t}$ = current, I

so $P = IV$

In other words, the electrical power converted in a device is given by the product of the current in it and the voltage across it. The units of electrical power are watts (W) – the same as mechanical power – as long as the current is in amperes and the p.d. is in volts.

Worked example

1 An X-ray tube operates at 50 kV. The maximum beam current is specified as 1.0 mA. What is the maximum safe power?

2 An electric iron is marked as 240 V, 1.8 kW. What is the current in the heater filament under normal operating conditions?

Answer

1 For safe power:

$P = IV = 1.0 \times 10^{-3}\,\text{A} \times 50 \times 10^{3}\,\text{V} = 50\,\text{W}$

2 From $P = IV$:

$I = \dfrac{P}{V} = \dfrac{1800\,\text{W}}{240\,\text{V}} = 7.5\,\text{A}$

Experiment

Investigating how the efficiency of an electric motor varies for different loads

This is an extension of the experiment measuring the power of a motor in Chapter 5.

Set up the arrangement shown in Figure 12.4. Adjust the power supply so that the motor can lift a mass of 100 g (that is, a force of 0.98 N) through a height of about 1 m in a few seconds (say 3–5 s).

Measure the time, t, that it takes to lift the mass, m, through a measured height, h, and record the corresponding p.d., V, and current, I. Put your results in Table 12.2.

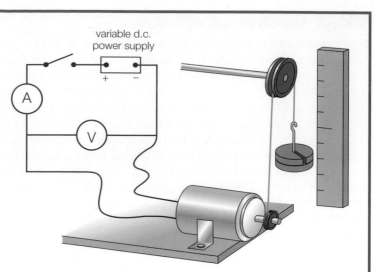

Figure 12.4 ▲

V/V	I/A	Power, IV/W	Force, mg/N	Height, h/m	Time, t/s	Power out, $\frac{mgh}{t}$ /W	Efficiency, $\frac{P_o}{P_i} \times 100\%$

Table 12.2 ◄

Repeat the experiment for further masses up to, for example, 500 g. Add your results to Table 12.2 and plot a graph of the efficiency against the force lifted.

A typical set of results is shown in Table 12.3.

V/V	I/A	Power, IV/W	Force, mg/N	Height, h/m	Time, t/s	Power out, $\frac{mgh}{t}$ /W	Efficiency, $\frac{P_o}{P_i} \times 100\%$
6.0	0.24	1.44	0.98	0.95	4.1	0.227	15.8
5.9	0.28		1.96	0.95	5.8		
5.8	0.31		2.94	0.95	7.5		
5.6	0.36		3.92	0.95	9.3		
5.4	0.40		4.90	0.95	11.6		

Table 12.3 ◄

Complete Table 12.3 and plot a graph of the efficiency of the motor for different loads (force). What deductions can you make from this graph? Do not worry if this graph differs significantly from the one you obtained in your experiment – different motors operating under varying conditions can produce very different results.

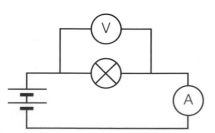

Figure 12.5 ▲

12.5 Electrical energy

In the circuit shown in Figure 12.5, electrical energy is converted into thermal and light energy in the filament of the torch bulb.

In the previous section we deduced that:

$$P = IV$$

If both sides of this equation are multiplied by time Δt:

$$P\Delta t = IV\Delta t \text{ or } \Delta W = IV\Delta t$$

where ΔW is the work done (or the energy converted) in time Δt.

Worked example

In the circuit shown in Figure 12.5, the current in the filament is 250 mA and the p.d. across the bulb is 2.2 V. How much electrical energy is converted into thermal energy and light in one minute?

Answer

$$\Delta W = IV\Delta t$$
$$= 0.25\,A \times 2.2\,V \times 60\,s$$
$$= 33\,J$$

In practice, a filament lamp converts more than 95% of electrical energy into thermal energy, so only about 1 J of light is produced in the above example. A filament lamp is therefore a very inefficient way of generating light.

This is why we are being encouraged to switch over to 'low-energy' compact fluorescent lamps (CFLs). Although a CFL costs more to manufacture than a filament lamp, it produces the same amount of light for less than 25% of the energy and also lasts about eight times longer (typically about 8000 hours compared with 1000 hours for an average filament lamp). It is estimated that if a CFL were to be on for as little as one hour per day, it would only take $2\frac{1}{2}$ years to recover the initial cost and that the savings over the lifetime of the lamp would amount to several times its original cost.

Furthermore, the benefit for the environment would be huge. Replacing a single filament lamp with a CFL would reduce emissions of carbon dioxide by 40 kg per year, and if each house in the UK were to install just three low-energy lamps, it would save enough energy to run the country's street lights for a year. This is why governments are encouraging the switch to low-energy lamps, and legislation may eventually lead to filament lamps becoming illegal.

REVIEW QUESTIONS

1. a) The SI base units for the watt are

   A $kg\,m\,s^{-2}$

   B $kg\,m^2\,s^{-2}$

   C $kg\,m\,s^{-3}$

   D $kg\,m^2\,s^{-3}$

   b) The SI base units for the volt are

   A $kg\,m^2\,s^{-2}\,A^{-1}$

   B $kg\,m^2\,s^{-3}\,A^{-1}$

   C $kg\,m^2\,s^{-3}\,A$

   D $kg\,m^2\,s^{-3}\,A$

2. The following quantities are often used in the study of electricity:

   charge   current   energy   potential difference   power

   Complete Table 12.4 by putting the appropriate quantity or quantities in the second column.

	Quantity or quantities
Which quantity is the product of two other quantities?	
Which quantity is one of the quantities divided by one of the other quantities? (3 possible answers).	
Which quantity, when divided by time gives another quantity in the table? (2 possible answers)	

Table 12.4 ▲

3. An electric iron is rated at '230 V, 1.5 kW'.

   a) What is the value of the current in the heating element?

   b) How much charge passes through the element during 20 minutes of use?

   c) How much thermal energy is produced during this time?

4 An experiment to investigate the efficiency of an electric motor is set up as shown in Figure 12.6.

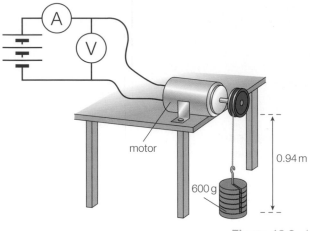

Figure 12.6 ▲

The mass of 600 g was lifted from the floor to a height of 0.94 m in 7.7 s and the meter readings were 3.2 V and 0.75 A.

a) How much work did the motor do lifting the mass?

b) What was the useful output power of the motor?

c) How much electrical power was being supplied by the cells?

d) Use your answers to parts b) and c) to show that the efficiency of the electric motor was about 30%.

e) Explain what happened to the rest of the energy.

5 A 1.0 m length of 32 swg copper wire is found to have a current of 1.8 A in it when the potential difference across it is 0.53 V.

a) Draw a circuit diagram to show how you would check these values.

b) Calculate the rate at which the power supply does work on the wire.

c) If copper has 8.0×10^{28} free electrons per cubic metre, and 32 swg wire has a cross-sectional area of 0.059 mm^2, calculate the drift speed of the electrons in the wire.

d) Use your answers to parts b) and c) to calculate the force exerted on the electrons in the wire.

6 The data in Table 12.5 are taken from the owner's manual of a car.

Alternator	14 V / 70 A
Starter motor	12 v / 1.5 kW
Battery	12 V / 62 A h
Maximum engine power	180 kW
Headlights (× 2)	12 V / 60 W
Sidelights (× 4)	12 V / 5 W
Spare fuses	8 A and 16 A
Mass	1740 kg (laden)

Table 12.5 ▲

a) How much power is delivered by the alternator?

b) How much current is taken by the starter motor?

c) The headlamps are fused individually. Explain which value of spare fuse you would use for a headlight.

d) Calculate how much energy can be produced by the battery (62 A h means the battery can provide a current of 1 A for 62 hours).

e) Show that nearly $\frac{1}{3}$ of the battery capacity is used up if the sidelights are accidentally left on for 12 hours overnight.

f) Show that only about $\frac{1}{4}$ of the maximum power that can be developed by the engine is used doing work against gravity when the car is driven at 90 km per hour up a gradient of 10%.

7 A manufacturer claims that replacing a 60 W filament lamp with an 18 W low-energy lamp, which provides the same amount of light, will save more than £40 over the lifetime of the lamp.

a) Use the following data to test this claim:
Cost of low energy lamp   £2
Estimated life of low energy lamp   8000 hours
Cost of a 'unit' of electricity   15p
(A 'unit' of electricity is 1 kW h, which is the energy used when a device rated at 1 kW is used for one hour.)

b) Discuss any other advantages of replacing filament lamps with low-energy lamps.

13 Current–potential difference relationships

In this section you will investigate how the current in different electrical components depends on the potential difference applied to the component.
You will learn that the **resistance** of a component is a measure of its opposition to the flow of charge through it and the importance of understanding the difference between resistance and **Ohm's law**.

13.1 Varying the potential difference and current in a circuit

In order to investigate how the current in an electrical component depends on the potential difference (p.d.) across it, we need to be able to vary the applied p.d. Some power supplies are continuously variable, but it is important that you understand how to set up a variable supply using a fixed power source and a variable resistor called a **rheostat**.

A rheostat can be used to either control the current in a circuit (Figure 13.1a) or give a continuously variable potential difference (Figure 13.1b). In the latter case, it is said to be a **potential divider**. Rheostats manufactured specifically for this purpose are called **potentiometers**; this term is rather misleading as 'meter' suggests that it is measuring something, which it is not! We will discuss potential dividers in more detail in Chapter 15.

Experiment

Set up the circuit shown in Figure 13.1a, taking care to connect the rheostat as shown. Use the rheostat to vary the current in the circuit. Record the minimum and maximum values of current and p.d. that can be achieved.

Now set up the circuit in Figure 13.1b. Look carefully at the way in which the rheostat is connected to act as a potential divider. Use the rheostat to vary the p.d. across the lamp.

What are the minimum and maximum values of p.d. and current with this circuit?

Which do you think is the more useful circuit?

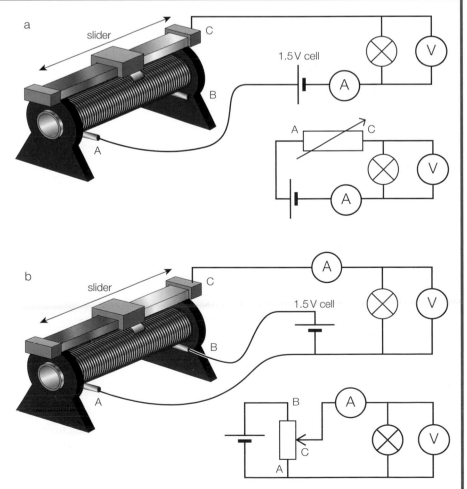

Figure 13.1 ▶

13.2 *I–V* characteristics for a metallic conductor

Cut just over a metre length of 32 swg nichrome wire (or the 0.25 mm metric equivalent) and tape it to a metre rule as shown in Figure 13.2. Set up a potential divider circuit to provide a variable p.d. and connect a 1.000 m length of the wire to the circuit by means of crocodile clips. **Ensure that the clips make a good connection.** Set the p.d. to its minimum value (that is, with the rheostat slider at the end nearer the negative contact) and switch on. Gradually increase the p.d. and record the values of p.d., *V*, and current, *I*, up to the maximum p.d. possible. Tabulate your values of *V* and *I*.

Repeat the experiment but with the terminals of the cell reversed. You should not change the polarity of the meters, which will now indicate negative values. The p.d. across the wire and the current in the wire are now in the opposite direction to that in the first experiment and should therefore be recorded as having negative values, as displayed on the meters.

Plot your data on a set of axes as shown in Figure 13.3. Note that it is conventional to plot *I* (on the *y* axis) against *V* (on the *x* axis).

A typical set of results is shown in Table 13.1.

V/V	0.00	0.50	1.00	1.50	2.00	2.50	3.00
I/mA	0	28	54	83	108	138	163

With cell terminals reversed

V/V	0.00	−0.50	−1.00	−1.50	−2.00	−2.50	−3.00
I/mA	0	−26	−56	−81	−110	−136	−165

Table 13.1 ▲

Plot the data on a suitable set of axes. What can you deduce about the relationship between *I* and *V* from your graph?

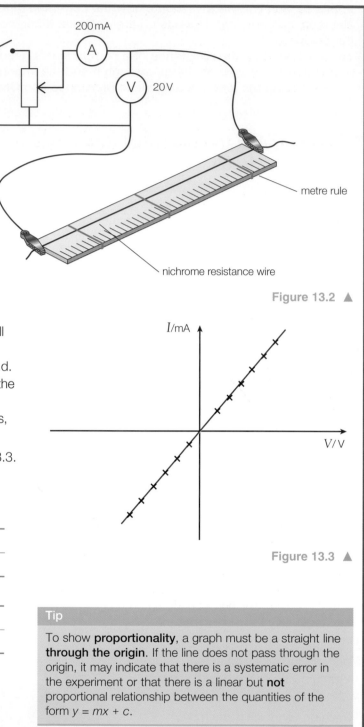

Figure 13.2 ▲

Figure 13.3 ▲

To show **proportionality**, a graph must be a straight line **through the origin**. If the line does not pass through the origin, it may indicate that there is a systematic error in the experiment or that there is a linear but **not** proportional relationship between the quantities of the form *y = mx + c*.

13.3 Ohm's law

From experiments similar to that above, Ohm discovered that **for metals at a constant temperature, the current in the metal is proportional to the potential difference across it**.

This is now known as **Ohm's law**, and any electrical component for which the current is proportional to the voltage is said to be 'ohmic'.

13.4 Resistance

The resistance (symbol R) of an electrical component is a measure of its opposition to an electric current flowing in it. This resistance is caused by collisions of the electrons with the vibrating lattice ions as the electrons 'drift' through the material of the conductor (see also $I = nAvq$ on page 122).

Resistance is **defined** by the equation:

$$R = \frac{V}{I}$$

This equation can be remembered by means of the diagram in Figure 13.4.

The **unit** of resistance is the ohm (Ω). **A conductor has a resistance of 1 Ω if a potential difference of 1 V across it produces a current of 1 A in the conductor.**

Figure 13.4 ▲

Tip

Take care – resistances are often quoted in kΩ or MΩ and currents in mA or µA, so always be careful to convert to base units.

Worked example

What is the potential difference across a 22 kΩ resistor when the current in it is 50 µA?

Answer

$$V = IR = 50 \times 10^{-6}\,\text{A} \times 22 \times 10^3\,\Omega$$
$$= 1.1\,\text{V}$$

Be careful to distinguish between **Ohm's law** and **resistance**. Ohm's law (Section 13.3) says that, under certain conditions, current is proportional to potential difference or, putting it another way, $\frac{V}{I}$ is constant.

This means that if a conductor obeys Ohm's law, **its resistance is constant** within the conditions specified – as in Figure 13.3 on page 134. In this instance, because the graph is of I against V, the resistance is the **inverse** of the gradient.

Worked example

1 Discuss whether the nichrome wire used in the experiment on page 134 in Section 13.2 obeys Ohm's law.

2 Use your graph to calculate the resistance of a 1.000 m length of nichrome wire.

Answer

1 As the graph is a straight line through the origin, the current is proportional to the potential difference, which indicates that the nichrome wire does obey Ohm's law.

2 The resistance of the wire is given by $R = \frac{V}{I}$.

As the graph is a straight line through the origin, its gradient is equal to $\frac{I}{V}$.

This means that the resistance can be found from the inverse of the gradient:

$$\text{Resistance} = \frac{V}{I} = \frac{1}{\text{gradient}}$$

$$= \frac{1}{0.0547\,\text{A V}^{-1}}$$

$$= 18.3\,\Omega$$

13.5 *I–V* characteristic for a tungsten filament lamp

Experiment

Set up the circuit as shown in Figure 13.5, with the voltage from the power supply set at its minimum value. Switch on and gradually increase the p.d., recording the p.d., *V*, and corresponding current, *I*, at regular intervals up to a maximum p.d. of 12 V.

Tabulate your results and **switch off**. Plot your results on a set of axes similar to those shown in Figure 13.3.

After at least five minutes have elapsed (Why?), reverse the power supply connections and repeat the experiment, remembering to start with the minimum p.d. as before. Add this data to your graph, which should now look like Figure 13.6.

What deductions can you make from the graph regarding the electrical characteristics of the filament?

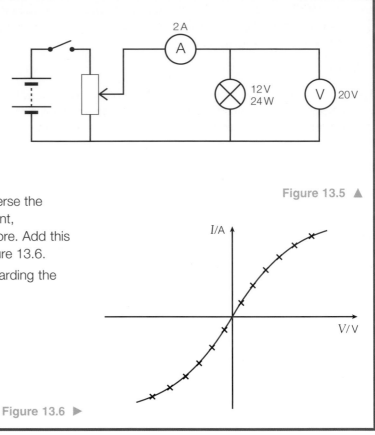

Figure 13.5 ▲

Figure 13.6 ▶

Worked example

A typical I–V graph for a tungsten filament lamp is shown in Figure 13.7.

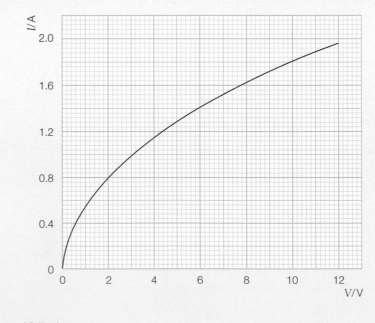

Figure 13.7 ▲

1 Discuss whether the lamp obeys Ohm's law.

2 The lamp from which the data was obtained was rated as 12 V, 24 W. What is the resistance of the filament when the lamp is operating under these conditions?

3 Use the graph to determine the resistance of the filament when:

a) the p.d. across it is 8.0 V

b) the current in it is 0.80 A.

Answer

1 As the graph of I against V is not a straight line through the origin, I is not proportional to V and so the lamp does not obey Ohm's law. This is because its temperature rises as the current increases.

2 Using $P = IV$:

$$I = \frac{P}{V} = \frac{24\,\text{W}}{12\,\text{V}} = 2.0\,\text{A}$$

$$R = \frac{V}{I} = \frac{12\,\text{V}}{2.0\,\text{A}} = 6.0\,\Omega$$

3 a) When the p.d. across the filament is 8.0 V, we can read off from the graph that the current is 1.62 A, so:

$$R = \frac{V}{I} = \frac{8.0\,\text{V}}{1.62\,\text{A}} = 4.9\,\Omega$$

b) When the current in the filament is 0.80 A, we can read off from the graph that the p.d. is 2.0 V, so:

$$R = \frac{V}{I} = \frac{2.0\,\text{V}}{0.80\,\text{A}} = 2.5\,\Omega$$

> **Tip**
>
> Remember that resistance can be found from the gradient of an I–V graph only if the graph is a straight line through the origin – that is, if Ohm's law is obeyed. In all other cases, R must be calculated from $R = \frac{V}{I}$ using discrete values of I and V.

13.6 *I–V* characteristic for a semiconductor diode

Experiment

Set up the circuit shown in Figure 13.8a – be careful to observe the polarity of the diode (see Figure 13.8b). The diode is said to be **forward biased** when connected this way around. Note that the 10 Ω resistor is included to limit the current in the diode and prevent it being damaged.

Switch on and very slowly increase the p.d. Carefully observe what happens to the current when you do this. Record values of I and V up to the maximum current obtainable.

Reverse the power supply connections so that the diode is **reverse biased** and repeat the experiment.

Plot all of your data on a suitable set of *I–V* axes. You should get a graph that looks like Figure 13.9. Note that the current in the diode when it is reverse biased seems to be zero. In reality, a very small current flows, but it is so small that it is not

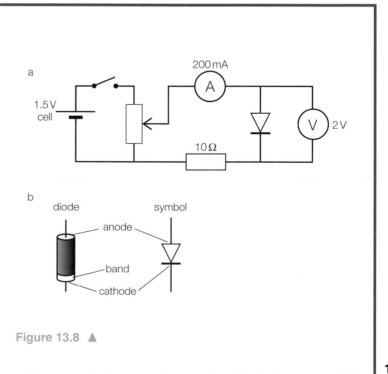

Figure 13.8 ▲

detected by the milliammeter. However, we normally think of a diode not conducting when it is reverse biased.

A typical set of results is shown in Table 13.2.

V/V	0.00	0.40	0.56	0.64	0.67	0.71	0.73	0.74
I/mA	0.0	0.0	1.0	5.0	10.0	20.0	30.0	40.0

With power supply terminals reversed

V/V	0.00	−0.20	−0.40	−0.60	−0.80	−1.00	−1.20	−1.40
I/mA	0.0	0.0	0.0	0.0	0.0	0.0	0.0	0.0

Table 13.2 ▲

1 Plot the data on a set of axes as shown in Figure 13.9.

2 Use your graph to determine:

a) the p.d. needed to make the diode conduct

b) the resistance of the diode when the current in it is 6.5 mA

c) the current in the diode when its resistance is 50 Ω. (Hint: construct a straight line on your graph corresponding to a resistance of 50 Ω.)

The answers to Question **2** should be:

a) about 0.45 V

b) 100 Ω

c) 14 mA.

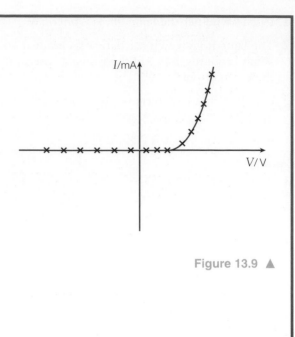

Figure 13.9 ▲

REVIEW QUESTIONS

1 Which of the graphs in Figure 13.10 could be a plot of:

a) current against potential difference for a semiconductor diode

b) resistance against potential difference for a tungsten filament lamp?

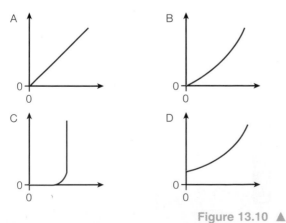

Figure 13.10 ▲

2 Graphs A and B in Figure 13.11 show the *I–V* characteristics for a carbon resistor of nominal value 33 Ω ±5% and a tungsten filament lamp rated 12 V, 5 W.

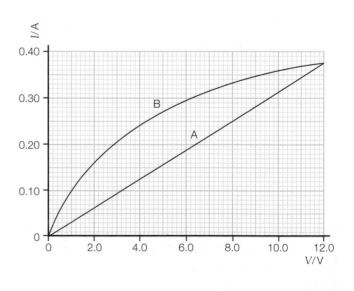

Figure 13.11 ▲

a) Explain the different shapes of the two graphs.

b) i) What value does graph A give for the resistance of the carbon resistor?

ii) Does this fall within the stated tolerance?

c) i) What value does graph B give for the power of the lamp when the potential difference across it is 12.0 V?

   ii) By what percentage does this differ from the stated value?

d) What is the resistance of the lamp when the p.d. across it is:

   i) 12.0 V   ii) 1.0 V

e) The resistor and lamp are connected in **series** with a 12 V power supply, as shown in Figure 13.12.

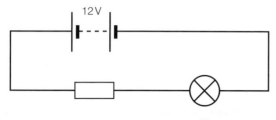

Figure 13.12 ▲

   i) Explain why the lamp glows dimly.

   ii) Discuss whether the resistor or the lamp consumes the greater power.

3 A student carried out an experiment on a 1.00 m length of nichrome ribbon taken from the heating element of an old toaster that was rated at 240 V, 1000 W. She found that the ribbon obeyed Ohm's law and had a resistance of 11.3 Ω when she applied potential differences of up to 12 V.

a) Define:

   i) Ohm's law

   ii) resistance.

b) What would be the value of the current in the wire when the potential difference across it was 2.0 V?

c) Sketch a graph of current against p.d. for values of p.d. up to 2.0 V, labelling the axes with appropriate values.

d) Explain, with a circuit diagram, how you could obtain this graph using a data logger. State one advantage and one disadvantage of using a data logger for an experiment such as this.

e) Estimate the length of ribbon making up the heating element of the toaster. What assumption do you have to make?

4 a) Sketch the *I–V* characteristic for a typical silicon diode.

b) A graph of the resistance of such a diode for different values of current is shown in Figure 13.13.

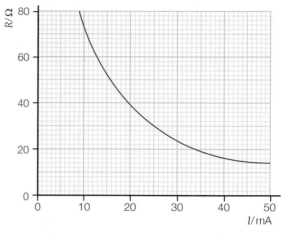

Figure 13.13 ▲

The circuit shown in Figure 13.14 is to be constructed so that the diode operates at a current as close to 25 mA as possible.

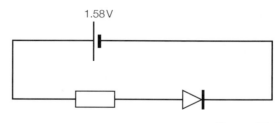

Figure 13.14 ▲

The following resistors are available: 10 Ω, 22 Ω, 33 Ω and 47 Ω. Which value would be the most suitable?

139

14 Resistance and resistivity

The concept of **resistance** was introduced in Section 13. We will now investigate the factors that determine the resistance of a conductor, in particular a property of the material from which the conductor is made, called its **resistivity**.
You will also see how the resistivity of metals and semiconductors depends on their temperature.

14.1 Resistance

It was convenient to introduce the concept of resistance in Chapter 13 when looking at the I–V characteristics for different electrical components. The resistance R of a component was defined as:

$$R = \frac{V}{I}$$

with V measured in volts, I in amperes and R therefore in ohms (Ω). The following worked example will act as a reminder.

Worked example

1 Explain why the graph in Figure 14.1a shows that the component obeys Ohm's law and calculate the resistance of the component.

2 Calculate the resistance of the component in Figure 14.1b when the current in it is $80\,\text{mA}$.

Answer

1 As the graph is a straight line **through the origin**, the current in the component is proportional to the potential difference across it, which is Ohm's law.

$$R = \frac{V}{I}$$

$$= \frac{6.0\,\text{V}}{120 \times 10^{-3}\,\text{A}}$$

$$= 50\,\Omega$$

2 Clearly the component does not obey Ohm's law, and as its resistance is not constant, we cannot use the gradient of the graph. We simply use the equation $R = \frac{V}{I}$.
Putting in the values for $I = 80\,\text{mA}$:

$$R = \frac{V}{I}$$

$$= \frac{0.80\,\text{V}}{80 \times 10^{-3}\,\text{A}}$$

$$= 10\,\Omega$$

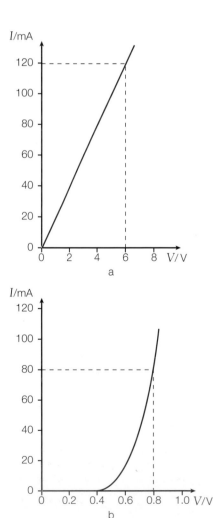

14.2 Power dissipation in a resistor

In Chapter 12, we derived that the power P transferred in an electrical component was given by:

$$P = IV$$

and that if I was in amperes and V in volts, then P would be in watts – the same unit as mechanical power.

Figure 14.1 ▲

Worked example

Show that the units of electrical power as defined by $P = IV$ are consistent with those for mechanical power, defined as the rate of doing work.

Answer

$$I = \frac{Q}{t} = C\,s^{-1} \text{ and } V = \frac{W}{Q} = J\,C^{-1}$$

$$\Rightarrow IV = C\,s^{-1} \times J\,C^{-1}$$

$$P = J\,s^{-1} = W$$

When a charge flows through a resistor, work is done on the resistor. It is sometimes convenient to express the power transferred in the resistor in terms of its resistance R and the current I in it.

We know that:

$$R = \frac{V}{I} \text{ or, rearranged, } V = IR, \text{ so:}$$

$$P = IV = I \times (IR) = I^2R$$

We can similarly show, by substituting $I = \frac{V}{R}$, that:

$$P = \frac{V^2}{R}$$

> **Tip**
>
> Learn that:
> $$P = IV = I^2R = \frac{V^2}{R}$$

Worked example

A manufacturer indicates that the maximum safe power for a particular range of resistors is 250 mW. What is the maximum safe current for a 47 kΩ resistor from this range?

Answer

$$P = I^2R$$

Rearranging:

$$I^2 = \frac{P}{R} = \frac{250 \times 10^{-3}\,W}{47 \times 10^3\,\Omega} = 5.32 \times 10^{-6}\,A^2$$

$$I = 2.3\,mA$$

We often say that power is **dissipated** in a resistor, particularly if that resistor is in the form of the filament of a lamp or the element of an iron or kettle. The word 'dissipated' means 'scattered'. The electrical energy transferred in the resistor increases the potential energy and the random kinetic energy of the atoms of the material of the resistor (what we call the **internal energy** of the atoms).

Worked example

An electric filament lamp is rated at 240 V, 60 W.

1   When it is operating under these conditions, what is:
   a) the current in the filament
   b) the resistance of the filament?

2   At the instant the lamp is switched on, its resistance is 80 Ω. At this instant, what is:
   a) the current in the filament
   b) the power dissipated in the filament?

Answer

1   **a)** $P = IV$, so $I = \dfrac{P}{V} = \dfrac{60\,W}{240\,V} = 0.25\,A$

$$\textbf{b)}\ R = \frac{V}{I} = \frac{240\,V}{0.25\,A} = 960\,\Omega$$

$$\textbf{2 a)}\ I = \frac{V}{R} = \frac{240\,V}{80\,\Omega} = 3\,A$$

$$\textbf{b)}\ P = IV = 3.0\,A \times 240\,V = 720\,W$$

The above example shows the large current (and power) 'surge' that occurs at the instant the lamp is switched on. This explains why the filament of a lamp heats up and reaches its operating temperature very quickly.

Experiment

The current surge when a lamp is switched on can be investigated using a data logger, which has the advantage of being able to take a large number of readings very quickly. Using the circuit shown in Figure 14.2, the current and potential difference can be measured for the first 200 ms after the lamp has been switched on.

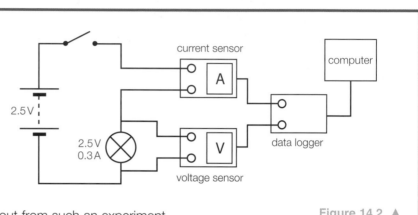

Figure 14.2 ▲

Figure 14.3 shows a typical computer print-out from such an experiment.
The data logger is switched on at $t = 0$ and the lamp is switched on at $t = 100\,ms$.

Use the print-out to determine:

1 the resistance of the filament when:

  a) it is operating normally

  b) the current in it is at its peak

2 the power developed in the filament when:

  a) it is operating normally

  b) the current in it is at its peak.

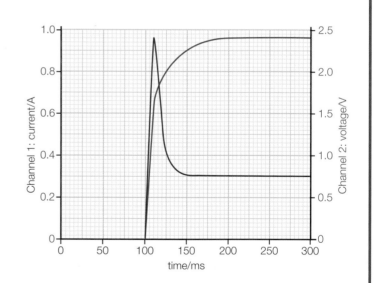

Figure 14.3 ▲

Answer

1 The resistance of the filament is given by $R = \frac{V}{I}$.

  a) When it is operating normally:

   $V = 2.40\,V$ (from the graph; 2.5 V nominally)

   $I = 0.30\,A$ (from the graph; 0.3 A nominally)

   $$R = \frac{2.4\,V}{0.30\,A} = 8.0\,\Omega$$

  b) When the current in it is at its peak:

   $V = 1.65\,V$ (from the graph)

   $I = 0.96\,A$ (from the graph)

   $$R = \frac{1.65\,V}{0.96\,A} = 1.7\,\Omega$$

2 The power developed in the filament is given by $P = VI$.

  a) When it is operating normally:

   $P = 2.4\,V \times 0.30\,A = 0.72\,W$

  b) When the current in it is at its peak:

   $P = 1.65\,V \times 0.96\,A = 1.58\,W$.

14.3 Resistivity

Experiment

The factors that affect the resistance of a wire can be investigated using the same 1 m length of 32 swg (diameter 0.2743 mm) nichrome wire taped to a metre rule, as in the experiment in Section 13.2 on page 134.

For the purpose of this experiment, it is more convenient to measure the resistance of the wire directly using an **ohmmeter** set on its 200 Ω range, as shown in Figure 14.4.

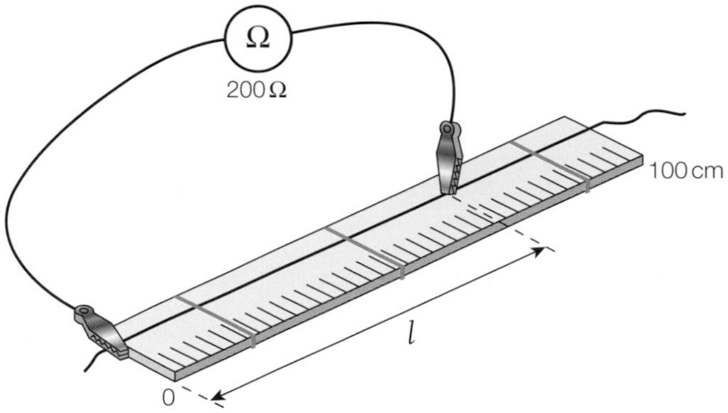

One crocodile clip is kept fixed at the zero end of the wire, while the other is pressed firmly on the wire to make contact at different lengths along the wire. The corresponding resistance, R, for each length, l, is recorded, and a graph of R against l is plotted. A typical set of results has been plotted in Figure 14.5.

The figure also shows the graph for a length of 30 swg (diameter 0.3150 mm) nichrome wire.

What can we deduce from these graphs?

Firstly, as they are both straight lines through the origin, it is obvious that:

$$R \propto l$$

The second deduction requires some calculation. The ratio of the areas of cross-section of the two wires is:

$$\frac{A_{30}}{A_{32}} = \frac{0.3150^2 \, \text{mm}^2}{0.2743^2 \, \text{mm}^2} = 1.32 \text{ (since } A = \frac{\pi d^2}{4})$$

The corresponding ratio of the resistances of a 1 m length of each wire is:

$$\frac{R_{30}}{R_{32}} = \frac{13.9 \, \Omega}{18.3 \, \Omega} = 0.76 = \frac{1}{1.32}$$

In other words:

$$R \propto \frac{1}{A}$$

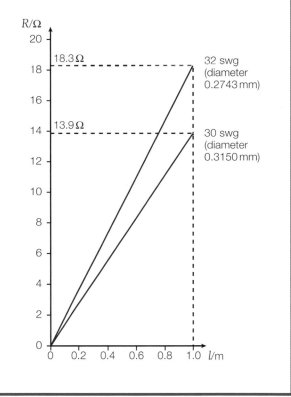

Figure 14.4 ▲

Figure 14.5 ▶

From the experiment, we can see that for a conductor (for example, a wire) of length l and uniform area of cross-section A:

$$R \propto l \text{ and } R \propto \frac{1}{A}$$

If we think about what is happening in the wire, this makes sense: if the wire is longer, it will be more difficult for the electrons to drift from one end to the other. If the wire has a larger cross-sectional area, it will be easier for the electrons to flow (like water in a larger diameter pipe).

Combining the two relationships:

$$R \propto \frac{l}{A} \text{ or } R = (constant) \times \frac{l}{A}$$

This constant is a property of the **material** of the wire, called its **resistivity**.

Resistivity has the symbol ρ ('rho'). The equation that defines resistivity is therefore:

$$R = \frac{\rho l}{A}$$

The **units** of resistivity can be derived by rearranging the equation:

$$\rho = \frac{RA}{l} = \frac{\Omega \, m^2}{m} = \Omega \, m$$

Worked example

Use the data for the 32 swg wire in Figure 14.5 to calculate a value for the resistivity of nichrome.

Answer

$$R = \frac{\rho l}{A}$$

where $A = \pi \left(\frac{d}{2}\right)^2$

Rearranging:

$$\rho = \frac{RA}{l}$$

Substituting $R = 18.3 \, \Omega$ when $l = 1.000 \, m$ and $d = 0.2743 \, mm$:

$$\rho = \frac{18.3 \, \Omega \times \pi \, (0.5 \times 0.2743 \times 10^{-3} \, m)^2}{1.000 \, m}$$

$$= 1.08 \times 10^{-6} \, \Omega \, m$$

> **Tip**
>
> Remember that:
> - resistivity is a property of the **material**
> - the unit of resistivity is ohms **times** metres **not** ohms per metre
> - in calculations, l must be in metres and A must be in m².

Worked example

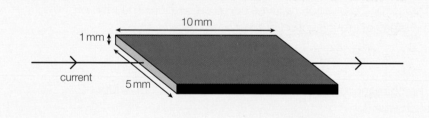

Figure 14.6 ▲

A carbon chip of resistivity $3.0 \times 10^{-5} \, \Omega \, m$ has the dimensions shown in Figure 14.6.

What resistance does the chip have for a current in the direction shown?

Answer

Using the formula:

$$R = \frac{\rho l}{A}$$

where $l = 10 \, mm = 10 \times 10^{-3} \, m$ and $A = 5 \, mm \times 1 \, mm$

$$= 5 \times 10^{-3} \, m \times 1 \times 10^{-3} \, m = 5 \times 10^{-6} \, m^2$$

$$R = \frac{3.0 \times 10^{-5} \, \Omega \, m \times 10 \times 10^{-3} \, m}{5 \times 10^{-6} \, m^2}$$

$$= 0.060 \, \Omega$$

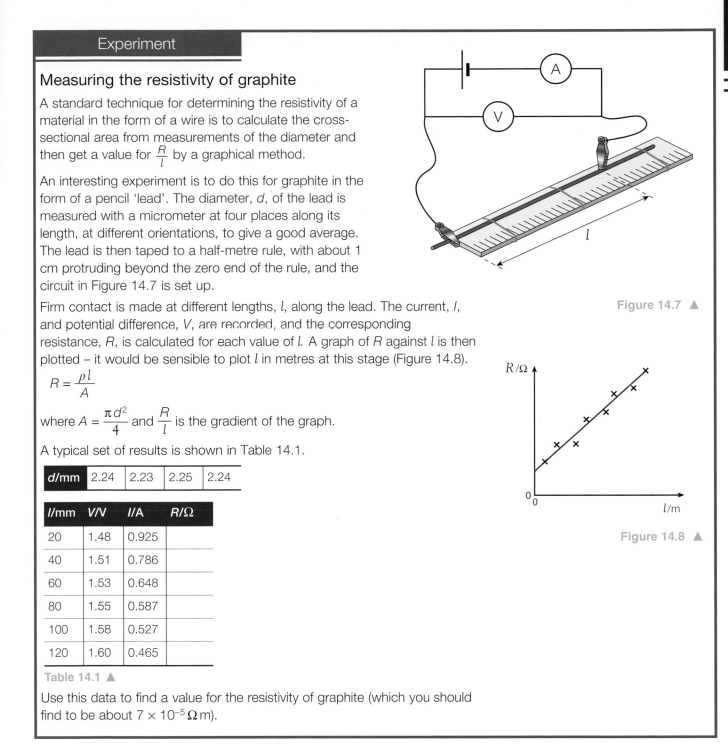

Experiment

Measuring the resistivity of graphite

A standard technique for determining the resistivity of a material in the form of a wire is to calculate the cross-sectional area from measurements of the diameter and then get a value for $\frac{R}{l}$ by a graphical method.

An interesting experiment is to do this for graphite in the form of a pencil 'lead'. The diameter, d, of the lead is measured with a micrometer at four places along its length, at different orientations, to give a good average. The lead is then taped to a half-metre rule, with about 1 cm protruding beyond the zero end of the rule, and the circuit in Figure 14.7 is set up.

Firm contact is made at different lengths, l, along the lead. The current, I, and potential difference, V, are recorded, and the corresponding resistance, R, is calculated for each value of l. A graph of R against l is then plotted – it would be sensible to plot l in metres at this stage (Figure 14.8).

$$R = \frac{\rho l}{A}$$

where $A = \frac{\pi d^2}{4}$ and $\frac{R}{l}$ is the gradient of the graph.

A typical set of results is shown in Table 14.1.

Figure 14.7 ▲

Figure 14.8 ▲

d/mm	2.24	2.23	2.25	2.24

l/mm	V/V	I/A	R/Ω
20	1.48	0.925	
40	1.51	0.786	
60	1.53	0.648	
80	1.55	0.587	
100	1.58	0.527	
120	1.60	0.465	

Table 14.1 ▲

Use this data to find a value for the resistivity of graphite (which you should find to be about $7 \times 10^{-5}\,\Omega\,\text{m}$).

The experiment shows the value of using a graphical technique for finding $\frac{R}{l}$.

The scatter of the points on the graph indicates that there is considerable **random error** – probably caused by variation in the pressure applied when making contact with the pencil lead or possibly by inconsistencies in the composition of the graphite. The fact that the graph does not go through the origin shows that there is also a significant **systematic error** caused by **contact resistance** due to poor contact between the crocodile clip and the pencil lead. A graphical method minimises the random error by averaging out the values and virtually eliminates the systematic error when the gradient is taken.

14.4 Effect of temperature on the resistivity of a metal

In Section 13.5, we looked at the *I–V* characteristics for a tungsten filament lamp and produced the graph shown in Figure 14.9.

Figure 14.9 ▶

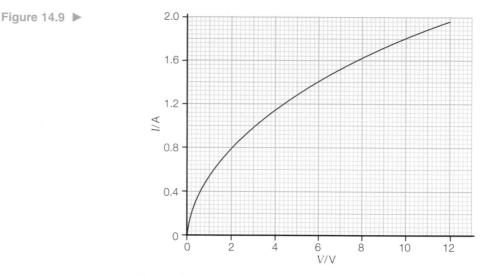

We said that the tungsten filament did not obey Ohm's law because the temperature of the filament rises as the current increases. We can infer from the graph that the resistance of the filament increases as its temperature rises.

Worked example

Use the graph in Figure 14.9 to calculate the resistance of the tungsten filament when the potential difference across it is 0.20 V, 6.0 V and 12.0 V.

Answer
Remember, we **cannot** use the gradient. We must read off the current, *I*, at each value of p.d., *V*, and then calculate the resistance from $R = \dfrac{V}{I}$.

From the graph: at $V = 0.20\,V$, $I = 0.20\,A$

$$\Rightarrow R = \frac{V}{I} = \frac{0.20\,V}{0.20\,A} = 1.0\,\Omega$$

You should check for yourself that the resistances at 6.0 V and 12.0 V are 4.3 Ω and 6.1 Ω, respectively.

The worked example shows that the resistance of the filament increases significantly as the temperature rises. The resistance of the filament at its normal operating temperature (about 3000 K) is some ten times greater than when it is 'off' – that is, at room temperature. This gives rise to a momentary 'surge' of current (and therefore power) when the lamp is first switched on, as shown in the experiment on page 142.

We can account qualitatively for the increase in the resistivity of metallic conductors with temperature by looking again at the 'free electron' model that we proposed for metals in Section 11.4. You will remember that we said that 'resistance' was caused by the vibrating positive ions in the crystal lattice of the metal impeding the flow of electrons. When the temperature of the metal is raised, the amplitude of vibration of the lattice ions increases and, as a result, there is an increased interaction between the lattice and the 'free' electrons.

In terms of the drift velocity equation, $I = nAvq$, *A* and *q* are constant for a given wire. For a metallic conductor, **n does not depend on the temperature** and so *n* is also constant. As the temperature rises, the increased vibrations of the lattice will reduce the drift velocity, *v*, of the electrons and so *I* will also decrease – that is, the **resistance increases with temperature**.

Experiment

The change in the resistance of a copper wire with temperature can be investigated with the arrangement shown in Figure 14.10.

A coil of copper should be formed by winding about 2 metres of enamelled 34 swg copper wire around a short length of plastic tube. About 2 cm of each end of the wire should be scraped clean to form a good electrical contact in the connector block.

As the resistance of the coil is less than 1 Ω, a voltmeter/ammeter method is needed to measure the resistance. Record the p.d. *V* and current *I* for a range of temperatures θ from room temperature up to the boiling point of the water. Put your results in Table 14.2.

θ/°C				
V/mV				
I/mA				
R/Ω				

Table 14.2 ◄

Plot a graph of *R* against θ. Start at 0 °C on the *x* axis and choose a scale on the *y* axis such that a value for R_0 – the resistance at 0 °C – can be read off. Your graph should look like Figure 14.11.

The graph shows that there is a linear increase of resistance with temperature. However, it is **not** a **proportional** relationship, as the graph does not go through the origin but has an intercept of R_0.

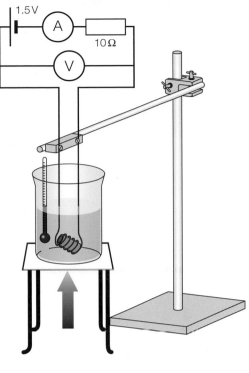

Figure 14.10 ▲

Exercise: Data analysis

Some typical results are shown in Table 14.3.

θ/°C	20	40	60	80	100
V/mV	114	122	130	137	144
I/mA	143	142	141	140	139
R/Ω					

Table 14.3 ◄

Complete the table by adding values for *R* and then plot a graph of *R* against θ, choosing your scales carefully as above.

There is a theoretical relationship (**which you do not have to know for examination purposes**) between the resistance and temperature of the form:

$$R = R_0 (1 + \alpha\theta)$$

where α is a constant – called the **temperature coefficient of resistivity** of copper. The equation can be rearranged to:

$$R = R_0\alpha\theta + R_0$$

The gradient is therefore equal to $R_0\alpha$ and the intercept on the *y* axis is R_0. Obtain values for the gradient and intercept and hence use this information to show that α has a value of about 4×10^{-3} K^{-1} for copper. As α is positive, a metal is said to have a positive temperature coefficient (PTC).

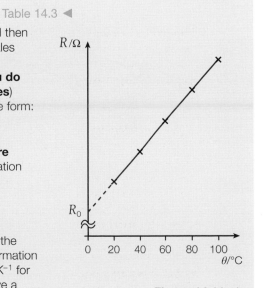

Figure 14.11 ▲

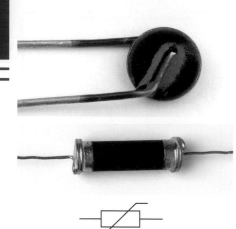

Metallic conductor	Semiconductor
n constant	n ↑↑
A constant	A constant
v ↓	v ↓
q constant	q constant
⇒	⇒
I ↓	I ↑
R ↑	R ↓

Table 14.4 ▲

14.5 Effect of temperature on the resistivity of a semiconductor

In order to demonstrate the effect of temperature on the resistance of a semiconductor, we can use a semiconducting device called a **thermistor**. The term comes from an abbreviation of 'thermal resistor', which, as its name suggests, is a resistor whose resistance is temperature dependent. Figure 14.12 shows two types of thermistor and the circuit symbol for a thermistor.

The dependence on temperature for a thermistor can be very simply shown by connecting a disc thermistor to an ohmmeter, holding it by its leads and noting its resistance at room temperature (typically about 20 °C). If it is then held firmly between the thumb and forefinger so that its temperature is your body temperature (about 37 °C), a very significant drop in the resistance of the thermistor should be observed. Such a thermistor, where the **resistivity decreases** with temperature, is said to have a **negative temperature coefficient** (NTC).

This can also be explained using $I = nAvq$. In a semiconductor, an increase in temperature can provide extra energy to release more charge carriers. This means that n **increases with temperature**. To a good approximation, n increases **exponentially** with the absolute temperature, which means that n shows a rapid increase as the temperature rises. Applying $I = nAvq$, A and q are constant as before but n increases by much more than the relatively small decrease in v. The overall effect is that I increases, so the **resistivity decreases with temperature**.

If the temperature is high enough, even some materials that we normally think of as **insulators** (for which n is very small) can begin to conduct. This is because the energy associated with the very high temperature breaks down the atomic structure so that more charge carriers are released.

A spectacular demonstration of this for glass is shown in Figure 14.13. **Note that in the interests of safety, this experiment should only be performed by your teacher!**

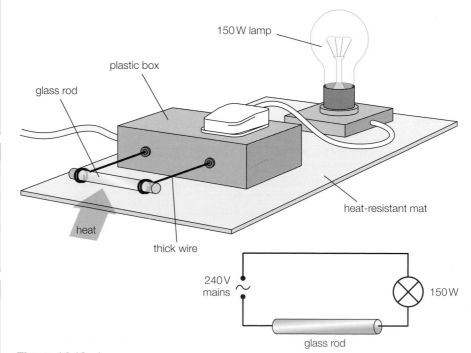

Figure 14.13 ▲

At room temperature, no current is observed as the glass rod acts as an insulator. When the rod is heated very strongly so that it glows red hot, the lamp begins to light and a current is registered by the ammeter – the glass has

Experiment

The effect of temperature on the resistance of a thermistor can be investigated using the apparatus shown in Figure 14.14.

Start with the thermistor immersed in crushed ice. Record the lowest steady temperature θ and the corresponding resistance R of the thermistor. Select the ohmmeter range that gives the most precise reading.

Now heat up the beaker of ice, being careful to keep the leads away from the flame and hot gauze. At intervals of about 10 °C, take away the Bunsen burner, stir thoroughly and record the temperature and corresponding resistance. Continue every 10 °C until the water boils. Put your results in Table 14.5.

$\theta/°C$									
R/Ω									

Table 14.5 ▲

Plot a graph of R against θ. You should get a curve like that in Figure 14.15.

This graph shows that the rate at which the resistance decreases is initially very rapid and then gradually becomes less.

Some typical results for a range of temperatures from 20 °C to 100 °C are recorded in Table 14.6.

$\theta/°C$	20	30	40	50	60	70	80	90	100
R/Ω	706	491	350	249	179	135	105	87	74

Table 14.6 ▲

Plot a graph of R against θ. What deductions can you make from the shape of the curve?

Extension

If you are familiar with logarithms and are studying physics beyond AS level, you might like to use this data to investigate whether the resistance of the thermistor is an exponential function of the temperature of the form:

$R = R_0e^{-k\theta}$

where R_0 is the resistance of the thermistor at 0 °C and k is a constant. Taking logarithms to base e, this equation becomes:

$\ln R = -k\theta + \ln R_0$

If the equation is valid, a graph of $\ln R$ against θ should be a straight line of gradient $-k$ and intercept $\ln R_0$. (You will find that the equation seems to be valid up to about 65 °C, with k approximately 0.034 K^{-1} and R_0 about 13 kΩ.)

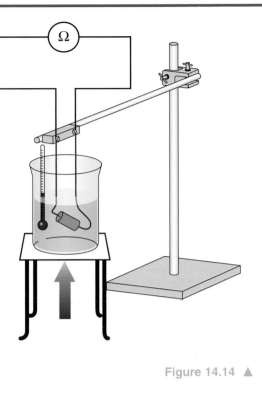

Figure 14.14 ▲

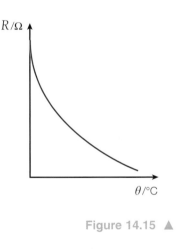

Figure 14.15 ▲

become a conductor! When the glass is really glowing, the source of heat can be removed, and the power dissipated by the current in the rod ($P = I^2R$) is sufficient to maintain the temperature of the glass and keep it conducting.

For the sake of completeness, it should be noted that semiconductors with a **positive temperature coefficient** (PTC) can also be constructed – their resistance actually increases with temperature. Indeed, it is even possible to make a semiconductor for which, over a reasonable temperature range, the increase in n is balanced by the decrease in v so that the resistance is constant! For example, the resistance of carbon resistors – of the type commonly found in school laboratories and in electronic circuits – remains more or less constant over a fairly wide range of temperatures.

Such a process is technically called 'doping', whereby a pure semiconductor such as germanium or silicon has atoms of an 'impurity' (such as arsenic or boron) added to it. The properties of the semiconductor created are very sensitive to the type and quantity of the impurity atoms.

REVIEW QUESTIONS

1 Which of the graphs in Figure 14.16 could be a plot against temperature of:

a) number of charge carriers per unit volume in a length of copper wire?

b) resistance of a length of copper wire?

c) resistance of a thermistor?

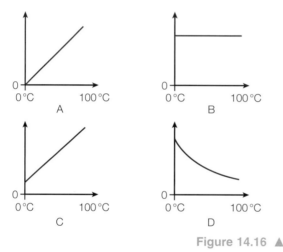

Figure 14.16 ▲

2 a) What is the resistance of the component represented in Figure 14.1b on page 140 when the current in it is 40 mA?

b) What p.d. would have to be applied across the component so that its resistance was $50\,\Omega$? (Hint: construct a straight line through the origin to represent a constant resistance of $50\,\Omega$ and see where it cuts the I–V curve for the component.)

3 For the carbon chip shown in Figure 14.6 on page 144, what is the resistance between:

a) the longer edges

b) the faces of the chip?

4 a) An electric kettle is rated at 240 V, 3.0 kW. Show that:

i) the current is such that it can operate safely with a 13 A fuse

ii) the resistance of the element when it is operating normally is about $19\,\Omega$.

b) If the kettle was operated on a 110 V supply and the

resistance of the element was assumed to remain constant, what would be:

i) the current in the element

ii) the power dissipated in the element?

c) Discuss whether the assumption that the resistance of the filament remains constant is reasonable.

5 a) Define **resistivity**.

b) Table 14.7 is taken from a data book.

Material	Gauge no.	Diameter/ mm	Area/ mm²	Ohms per metre	Resistivity/ Ωm
Copper	24	0.5588	0.2453	i)	1.72×10^{-8}
Constantan	28	ii)	0.1110	4.41	iii)
Nichrome	32	0.2743	iv)	18.3	v)

Table 14.7 ▲

Calculate the missing data i)–v).

6 The graph in Figure 14.17 shows data from an experiment to investigate how the current, I, varied with the p.d., V, across a 2.00 m length of nichrome wire of diameter 0.25 mm.

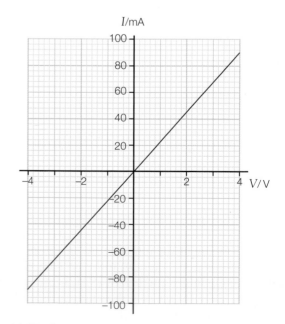

Figure 14.17 ▲

a) Explain why the graph shows that the wire obeys Ohm's law.

b) Use the data to calculate a value for the resistivity of nichrome.

7 a) Sketch a graph to show how the current in a 12 V, 5 W tungsten filament bulb depends on the potential difference across it.

b) Calculate the theoretical resistance of the filament when it is at its operating temperature.

c) Explain, with reference to the equation $I = nAvq$, why the resistance of the filament is much less when the lamp is 'off'.

8 The graph in Figure 14.18 shows the current–voltage characteristic for a thermistor.

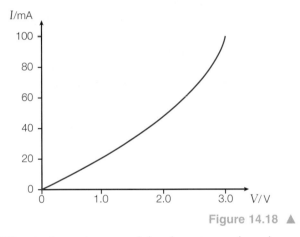

Figure 14.18 ▲

a) What is the resistance of the thermistor when the p.d. across it is:

i) 1.0 V

ii) 3.0 V?

b) Calculate the rate at which energy is being converted in the thermistor at each of these voltages.

c) i) What does your answer to part **b)** tell you about the temperature of the thermistor when the p.d. across it is 3 V compared with when the p.d. is 1 V?

ii) Explain, in terms of the equation $I = nAvq$, why the resistance changes in the way that it does in part **a)**.

15 Electric circuits

This is a very practical section in which you will carry out a number of experiments to investigate the behaviour of various electric circuits. You will learn how resistors can be combined in series and in parallel and how to calculate current, potential difference and power in circuits.

Conservation laws are fundamental concepts in physics. You should understand how the underlying principles of **conservation of energy** and **conservation of charge** apply to electric circuits.

15.1 Conservation of energy in circuits

Conservation of energy – that is, **energy cannot be created or destroyed but merely changed from one form to another** – is a fundamental concept of physics. Consider the circuit in Figure 15.1.

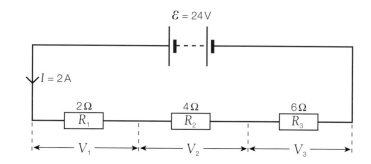

Figure 15.1 ▲

By the principle of conservation of energy, when a charge, Q, flows round the circuit:

energy converted by battery = energy dissipated in the three resistors
$$Q\mathcal{E} = QV_1 + QV_2 + QV_3$$
$$\mathcal{E} = V_1 + V_2 + V_3$$

As $V = IR$:
$$\mathcal{E} = IR_1 + IR_2 + IR_3$$

or circuit e.m.f.
$$\mathcal{E} = \Sigma IR$$

Tip

The symbol Σ ('sigma') means 'sum of'.

If 1 C of charge were to move round the circuit:

energy gained by charge in battery	=	energy lost by charge in R_1	+	energy lost by charge in R_2	+	energy lost by charge in R_3
(24 J)		(4 J)		(8 J)		(12 J)

Worked example

The circuit shown in Figure 15.1 is left on for 2.0 minutes.

1 Calculate:

  a) how much electrical energy is converted in the cell

  b) how much energy is dissipated in each of the resistors.

2 Explain how your answers show that energy is conserved.

Answer

1 a) Charge flowing $Q = It = 2\,A \times 120\,s = 240\,C$

Energy converted in cell $= \mathcal{E}Q = 24\,V \times 240\,C = 5.76\,kJ$

   b) Energy dissipated in $R_1 = I^2 R_1 t = (2\,A)^2 \times 2\,\Omega \times 120\,s = 0.96\,kJ$
Energy dissipated in $R_2 = I^2 R_2 t = (2\,A)^2 \times 4\,\Omega \times 120\,s = 1.92\,kJ$
Energy dissipated in $R_3 = I^2 R_3 t = (2\,A)^2 \times 6\,\Omega \times 120\,s = 2.88\,kJ$

2 Add up the energy dissipated in each of the resistors:

$0.96\,kJ + 1.92\,kJ + 2.88\,kJ = 5.76\,kJ =$ energy converted in cell

This shows that energy has been conserved.

15.2 Internal resistance

Unfortunately, not all of the chemical energy converted to electrical energy inside a cell emerges at the terminals of the cell. When a charge flows and produces a current in the cell, some of the energy is used up in 'pushing' the electrons through the cell. In other words, it is used to overcome the **internal resistance** of the cell, which is usually given the symbol **r**.

Consider a cell of e.m.f., $\mathcal{E}$, and internal resistance, r, which is connected to a lamp of resistance, R, as in Figure 15.2.

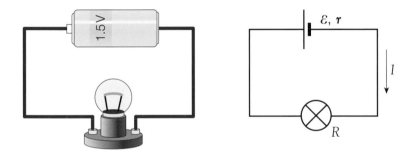

Figure 15.2 ▲

If there is a current, I, in the circuit:

rate of energy converted in cell	=	rate of work done against internal resistance	+	rate of work done lighting lamp
$\mathcal{E}I$	=	$I^2 r$	+	$I^2 R$
$\mathcal{E}$	=	Ir	+	IR

Rearranging: $IR = \mathcal{E} - Ir$

and putting $IR = V$: **$V = \mathcal{E} - Ir$**

This is effectively an application of the relationship $\mathcal{E} = \Sigma IR$ that we derived in Section 15.1.

The difference between the e.m.f., $\mathcal{E}$, of a cell and the potential difference, V, at its terminals is $\mathcal{E} - V = Ir$ and is sometimes called the '**lost volts**'. (Why do you think it is called this?)

An alternative rearrangement of the energy equation that enables us to find the current is:

$$I = \frac{\mathcal{E}}{R + r}$$

Worked example

A torch battery of e.m.f. 4.5 V and internal resistance 0.4 Ω is connected across a lamp of resistance 6.4 Ω.

1 What is the current in the lamp?

2 How much power is:
   a) dissipated in the lamp
   b) wasted in the cell?

Answer

1 Circuit current $I = \dfrac{\mathcal{E}}{R + r} = \dfrac{4.5\,V}{6.4\,\Omega + 0.4\,\Omega}$

$\qquad\qquad\quad = 0.66\,A$

2 a) For power dissipated in lamp:
   $P = I^2 R = (0.66\,A)^2 \times 6.4\,\Omega$
   $\qquad = 2.8\,W$

   b) For power wasted in cell:
   $P = I^2 r = (0.66\,A)^2 \times 0.4\,\Omega$
   $\qquad = 0.2\,W$

> **Tip**
>
> Remember:
>
> $V = \mathcal{E} - Ir$
>
> and
>
> $I = \dfrac{\mathcal{E}}{R + r}$

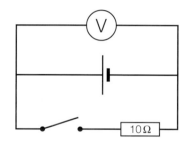

Figure 15.3 ▲

Worked example

In the circuit in Figure 15.3, the high-resistance voltmeter reads 1.55 V when the switch is open and 1.49 V when the switch is closed.

1 Explain why:
   a) the e.m.f. of the cell can be considered to be 1.55 V
   b) the voltmeter reading drops when the switch is closed.

2 Calculate the internal resistance of the cell.

Answer

1 a) As the voltmeter has a very high resistance, it takes virtually no current. Therefore, with the switch open, there is negligible current in the cell, so the reading of 1.55 V can be taken as its e.m.f.

   b) When the switch is closed, the 10 Ω resistor is brought into the circuit. This causes a current, I, in the circuit so that the potential difference, V, across the cell drops to $V = \mathcal{E} - Ir$. (With the switch open, $I = 0$, so that $V = \mathcal{E}$.)

2 From $V = \mathcal{E} - Ir$:

   $Ir = \mathcal{E} - V$
   $\quad = 1.55\,V - 1.49\,V$
   $\quad = 0.06\,V$

   For the 10 Ω resistor:

   $I = \dfrac{V}{R} = \dfrac{1.49\,V}{10\,\Omega}$
   $\qquad = 0.149\,A$

   $r = \dfrac{\mathcal{E} - V}{I} = \dfrac{0.06\,V}{0.149\,A}$

   $r = 0.40\,\Omega$

Finding the e.m.f. and internal resistance for a cell

Figure 15.4 shows a standard circuit to find the e.m.f. and internal resistance for a zinc-carbon cell.

Starting with the variable resistor (rheostat) at its highest value (to minimise any heating effects), record the current, I, in the cell and the potential difference, V, across its terminals for different settings of the rheostat.

Rearranging $V = \varepsilon - Ir$:

$$V = -Ir + \varepsilon$$

If a graph of V against I is plotted, we would expect to get a straight line of gradient $-r$ and intercept ε on the y axis (Figure 15.5). In practice, the line may not be straight because the internal resistance may not be constant – particularly for large currents.

A typical set of observations is recorded in Table 15.1. Plot a graph of this data and hence determine values for the e.m.f. and internal resistance of the cell. (You should get values of about 1.56 V and 0.60 Ω, respectively.)

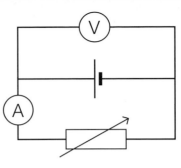

`Figure 15.4 ▲

I/A	0.20	0.40	0.60	0.80	1.00	1.20	1.40	1.60
V/V	1.44	1.32	1.20	1.09	0.95	0.84	0.73	0.59

Table 15.1 ▲

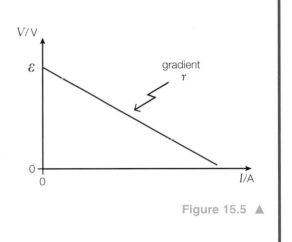

Figure 15.5 ▲

All power supplies – not just cells – have internal resistance, and the value of this internal resistance is often critical for the correct functioning of the power supply. For example, a car battery must have a **very small** internal resistance (typically as low as 0.01 Ω) because the starter motor takes a very large current when the engine is started (perhaps as much as 200 A). This is illustrated in the worked example.

On the other hand, a laboratory high voltage (E.H.T.) supply, for safety reasons, has a **very large** built-in internal resistance – typically 50 MΩ. This limits the current to a fraction of a milliamp, which is nevertheless still enough to give a nasty shock! E.H.T. supplies must therefore be treated with great care.

Worked example

1 A 12 V car battery has an internal resistance of 0.01 Ω.
   a) What is the potential difference across its terminals when the engine is started if the starter motor takes an initial current of 200 A?
   b) Explain why, if the driver has the headlights on, they are likely to go dim when he starts up the engine.
   c) Calculate how much power the battery delivers to the starter motor.

2 A mechanic of resistance 10 kΩ – as measured between his hand and the ground – accidentally touches the 'live' terminal of the battery and experiences a small electric shock. Calculate the power dissipated in his body.

Answer

1 **a)** Using $V = \varepsilon - Ir$:
$$V = 12\,V - (200\,A \times 0.01\,\Omega)$$
$$= 12\,V - 2\,V$$
$$= 10\,V$$

b) The headlights operate at full brightness when the p.d. across them is 12 V. If this is reduced to only 10 V when the engine is started, the headlights will go dim.

c) The power delivered by the battery to the starter motor is given by:
$$P = IV = 200\,A \times 10\,V$$
$$= 2000\,W \text{ (or } 2\,kW)$$

2 As the internal resistance of the battery is so much less than the resistance of the mechanic, we can ignore any effect of the internal resistance and simply assume that the current in the mechanic is given by
$$I = \frac{\varepsilon}{R} = \frac{12\,V}{10 \times 10^3\,\Omega} = 1.2 \times 10^{-3}\,A$$

$$P = IV = 1.2 \times 10^{-3}\,A \times 12\,V$$
$$= 0.014\,W \text{ (14 mW)}$$

15.3 Solar cells

Solar, or photovoltaic, cells are now widely used as a source of electric power. Your calculator is almost certainly powered by a solar cell, and you have probably seen solar cells being used to power road signs. Arrays of solar cells are used to power the electrical systems of satellites and can be used to supply electrical energy to buildings. These must not be confused with solar **heating** panels, in which the energy from the Sun is used to heat water flowing in an array of pipes in the panel.

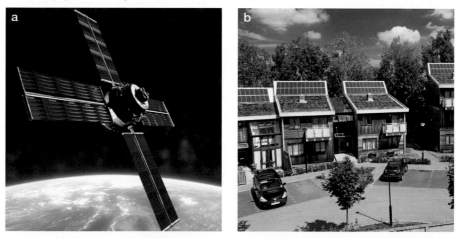

Figure 15.6 ▲
a) A satellite using solar cells to power its electrical systems and b) homes with solar cells on their roofs

When light strikes the photocell, it gives some of its energy to free electrons in the semiconductor material of the cell (commonly silicon). An electric field within the cell provides a force on the electrons. The electron flow provides the **current** and the cell's electric field causes a **voltage**. With both current and voltage, we have **power**, which is the product of the two.

The characteristics of a solar cell are dependent on the illumination. They can be investigated using the circuit shown in Figure 15.7. The value of the potentiometer required will depend on the type of cell and the amount of illumination.

Worked example

The data in Table 15.2 were obtained for the output of a solar cell using the circuit shown in Figure 15.7 when the cell was illuminated by normal laboratory light.

I/mA	0.010	0.015	0.020	0.025	0.030	0.035	0.040	0.045
V/V	2.07	1.97	1.84	1.72	1.60	1.44	1.28	1.10
R/kΩ								
P/μW								

Table 15.2 ▲

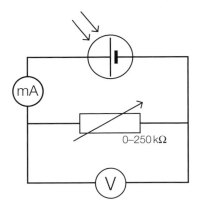

Figure 15.7 ▲

1 Complete Table 15.2 by adding values for the resistance across the cell, $R = \dfrac{V}{I}$, and the power generated, $P = VI$.

2 a) Draw a graph of *V* against *I* and comment on the shape of your graph.
   b) Use your graph to determine values for:
      i) the e.m.f. of the cell
      ii) the internal resistance of the cell for low current values.
   c) Plot a graph of *P* against *R* and use it to determine the maximum power generated.

Answer

1
I/mA	0.010	0.015	0.020	0.025	0.030	0.035	0.040	0.045
V/V	2.07	1.97	1.84	1.72	1.60	1.44	1.28	1.10
R/kΩ	207	131	92	69	53	41	32	24
P/μW	21	30	37	43	48	50	51	49

Table 15.3 ▲

2 a) The graph of *V* against *I* should look like Figure 15.8.

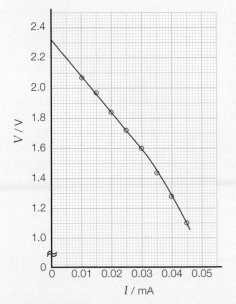

Figure 15.8 ▲

The graph is initially a straight line, which shows that the cell has a constant internal resistance for low current values. At larger current values, the internal resistance increases significantly.

b) **i)** The e.m.f. of the cell is given by the intercept when $I = 0$ – that is, e.m.f. = 2.3 V.

ii) The internal resistance of the cell for low current values is equal to the gradient of the straight part of the line. You can determine this for yourself and show that it is about 24 kΩ.

c) If you plot a graph of P against R, you will find that the maximum power generated is about 52 μW.

Review Question 6 on page 170 provides data for this cell when it is illuminated by a bench lamp. You will find that under these conditions, the e.m.f. of the cell increases to about 6.5 V and its internal resistance drops to about 3.6 kΩ. The maximum power generated is **much** greater – just over 2 mW.

15.4 Measuring the resistance of a component

We saw in Section 13.4 that resistance was defined as $R = \frac{V}{I}$.

This obviously gives a way of measuring resistance, as in the circuit shown in Figure 15.9.

The current, I, in the resistor is measured by the ammeter, the potential difference, V, across the resistor is measured with the voltmeter, and then $R = \frac{V}{I}$.

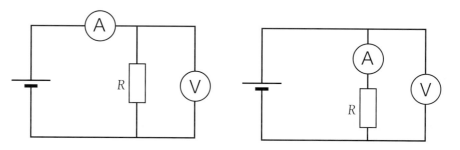

Figure 15.9 ▲   Figure 15.10 ▲

We saw in Section 12.2 that a small amount of current is taken by the voltmeter, so the current recorded by the ammeter is not, strictly speaking, the actual current in the resistor. As long as the voltmeter has a high resistance (or much greater than that of the resistor) we can ignore this effect. Alternatively, we could set up a circuit as shown in Figure 15.10.

We saw in Section 11.3 that ammeters should have a very small resistance, so they do not affect the current that they are measuring. Nevertheless, the ammeter might have an appreciable effect in the circuit of Figure 15.10, particularly if the resistor also has a low resistance. The ammeter **does**, indeed, measure the current in the resistor, but now the voltmeter measures the p.d. across both the resistor **and** the ammeter. In most cases the circuit shown in Figure 15.9 is preferable.

The voltmeter/ammeter method is rather cumbersome, however, as it requires two meters and a power supply, together with a calculation, to find the resistance. A quicker and easier way is to use a **digital ohmmeter**. As its name suggests, this instrument measures 'ohms' – that is, resistance – directly. Most digital multimeters have an 'ohms' range, or digital ohmmeters can be purchased as separate instruments.

You should understand that a digital ohmmeter does not actually measure resistance directly like an ammeter measures current. When you turn the dial on a multimeter to the 'ohms' range, you bring a battery into play, which produces a very small current in the component whose resistance is being determined. This current is measured by the meter and then converted into a resistance reading by the 'electronics' inside the meter. The general principle of an ohmmeter can be demonstrated by the experiment on page 162, in which a graphical method, rather than electronics, is used to find a resistance value.

Unfortunately, most school laboratory ohmmeters have a lowest range of $200\,\Omega$ (well $199.9\,\Omega$ actually!), so they can only measure to a precision of $0.1\,\Omega$. This means that for accurate measurements of resistance in the order of an ohm or less, we have to revert back to the voltmeter/ammeter method.

15.5 Resistors in series

Consider the circuit shown in Figure 15.11. If we were to insert an ammeter in the positions A, B, C and D in turn, we would observe that the current, I, was the same at each position. This arises as a result of the **conservation of charge** (Section 11.3) – in simple terms, the rate at which electrons leave each resistor must be the same as the rate at which they enter, as there is nowhere else for them to go!

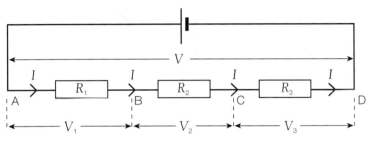

Figure 15.11 ◄

By connecting a voltmeter across AD and then across each of AB, BC and CD in turn, we would find that:

$$V = V_1 + V_2 + V_3$$

We have seen this result before in Section 15.1, where we discussed how it arises as a result of the application of the **conservation of energy**.

If the three resistors have a combined resistance R, then by using $V = IR$:

$$IR = IR_1 + IR_2 + IR_3$$

As I is the same throughout, **in series**:

$$R = R_1 + R_2 + R_3$$

15.6 Resistors in parallel

Now consider the parallel arrangement in Figure 15.12.

Figure 15.12 ◄

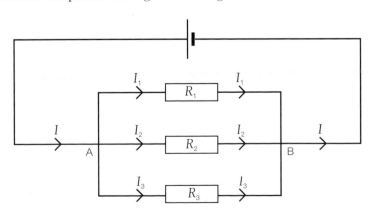

As the three resistors are connected to the common points A and B, the potential difference V (which could be measured by connecting a voltmeter between A and B) must be the same across each resistor.

We saw in Section 11.3 that from the conservation of charge at a junction:

$$I = I_1 + I_2 + I_3$$

where I is the total circuit current and I_1, I_2 and I_3 are the currents in R_1, R_2 and R_3, respectively.

If the three resistors have a combined resistance R, using $I = \frac{V}{R}$ we have:

$$\frac{V}{R} = \frac{V}{R_1} + \frac{V}{R_2} + \frac{V}{R_3}$$

As V is the same across each resistor, **in parallel**:

$$\frac{1}{R} = \frac{1}{R_1} + \frac{1}{R_2} + \frac{1}{R_3}$$

15.7 Series and parallel combinations

Problems involving series and parallel combinations of resistors should be tackled by combining the parallel resistors to find the equivalent single resistance and then adding up the series resistances.

Consider the circuit in Figure 15.13.

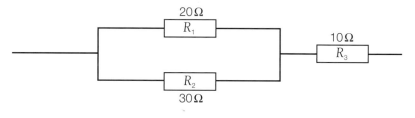

Figure 15.13 ▲

For the parallel arrangement of R_1 and R_2:

$$\frac{1}{R} = \frac{1}{R_1} + \frac{1}{R_2}$$

$$\frac{1}{R} = \frac{1}{20\,\Omega} + \frac{1}{30\,\Omega}$$

$$= 0.050\,\Omega^{-1} + 0.033\,\Omega^{-1} = 0.083\,\Omega^{-1}$$

$$R = \frac{1}{0.083\,\Omega^{-1}} = 12\,\Omega$$

We can now add this to R_3 to give the resistance of the combination of R_1, R_2 and R_3 as $10\,\Omega + 12\,\Omega = 22\,\Omega$. We can represent this calculation diagrammatically as in Figure 15.14.

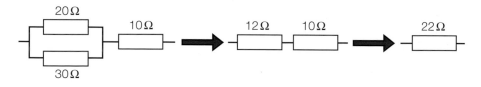

Figure 15.14 ▲

Worked example

Draw each arrangement that can be obtained by using three $10\,\Omega$ resistors either singly or combined in series and parallel arrangements (seven arrangements are possible!). Calculate all possible resistance values.

Answer
Figure 15.15 shows how the seven possible combinations can be obtained.

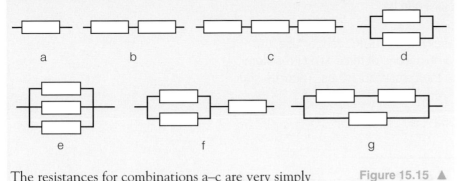

a   b   c   d

e   f   g

Figure 15.15 ▲

The resistances for combinations a–c are very simply
10 Ω, 20 Ω and 30 Ω, respectively.

For combination d:

$$\frac{1}{R} = \frac{1}{10\,\Omega} + \frac{1}{10\,\Omega}$$
$$= 0.10\,\Omega^{-1} + 0.10\,\Omega^{-1} = 0.20\,\Omega^{-1}$$
$$R = \frac{1}{0.20\,\Omega^{-1}} = 5.0\,\Omega$$

Similarly, for combination e:

$$\frac{1}{R} = \frac{1}{10\,\Omega} + \frac{1}{10\,\Omega} + \frac{1}{10\,\Omega}$$
$$= 0.1\,\Omega^{-1} + 0.1\,\Omega^{-1} + 0.1\,\Omega^{-1} = 0.3\,\Omega^{-1}$$
$$R = \frac{1}{0.30\,\Omega^{-1}} = 3.3\,\Omega$$

To find the resistance of combination f we need to combine the two parallel
resistors, which, from combination d, gives 5.0 Ω, and then add to this to the
third resistor to give:

$$R = 5.0\,\Omega + 10.0\,\Omega = 15.0\,\Omega$$

This is shown in Figure 15.16.

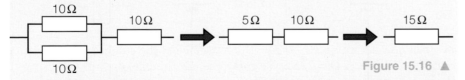

10 Ω   10 Ω   5 Ω   10 Ω   15 Ω
10 Ω

Figure 15.16 ▲

To find the resistance of combination g we need to add the two series
resistors first, which gives 20 Ω. There is then, effectively, 20 Ω in parallel
with the other 10 Ω resistor (Figure 15.17), so that:

$$\frac{1}{R} = \frac{1}{20\,\Omega} + \frac{1}{10\,\Omega}$$
$$= 0.05\,\Omega^{-1} + 0.10\,\Omega^{-1} = 0.15\,\Omega^{-1}$$
$$R = \frac{1}{0.15\,\Omega^{-1}} = 6.7\,\Omega$$

This is shown in Figure 15.17.

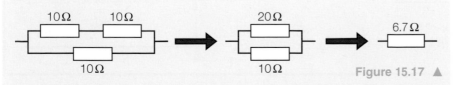

10 Ω   10 Ω   20 Ω   6.7 Ω
10 Ω   10 Ω

Figure 15.17 ▲

> **Tip**
>
> Remember:
> - the resistance of a combination of parallel resistors is always less than that of any of the individual resistors
> - for two **equal** resistors in parallel, the combined resistance is **half** that of each single resistor
> - always work out parallel combinations **separately** and add series values afterwards – **do not** try to do it all in one go!

Experiment

Series and parallel combinations of resistors can be investigated easily using an ohmmeter.

Set a digital multimeter (or ohmmeter) initially to its 200 kΩ range. Use it to measure the resistance of all possible combinations of three 10 kΩ resistors and compare your measured values with the calculated values. Tabulate your results as in Table 15.4.

Calculated resistance/kΩ	10.0	20.0	30.0	5.0	etc
Measured resistance/kΩ					

Table 15.4 ▲

You can use the worked example on page 161 to help you work out the various combinations and the calculated values, some of which are already given in Table 15.4. It would be advisable to change the range of the ohmmeter to 20 kΩ for the smaller resistance values (unless you have an 'auto-ranging' instrument).

The following experiment shows, in broad terms, the principle of an ohmmeter. It uses a graphical calibration to find an 'unknown' resistance (in this case, the resistance of a series/parallel combination) rather than 'electronics' to convert the measured current into a resistance reading. Such an experiment is typical of the type of experiment used for practical assessment. It can be set up very easily using a proprietary circuit board.

Experiment

Set up the circuit shown in Figure 15.18a. The circuit current, I, recorded by the ammeter is the current in the combination of the three resistors in series – that is, a resistance $R = 1.0\,k\Omega + 2.2\,k\Omega + 4.7\,k\Omega = 7.9\,k\Omega$.

Now connect a spare lead across the 1.0 kΩ resistor as shown in Figure 15.18b. We say that we have 'short-circuited' the 1.0 kΩ resistor. In effect, we have built a 'bypass' for the current so that there is no current (or at least a negligible amount) in the 1.0 kΩ resistor. The circuit resistance is now just $R = 2.2\,k\Omega + 4.7\,k\Omega = 6.9\,k\Omega$.

Repeat this to get values of resistance for all possible series combinations of the resistors, as shown in Table 15.5. Be careful not to short circuit all three resistors at once.

R/kΩ	7.9	6.9	5.7	4.7	3.3	2.2	1.0
I/mA							

Table 15.5 ▲

Plot a graph of R against I. You should get a curve like that in Figure 15.19.

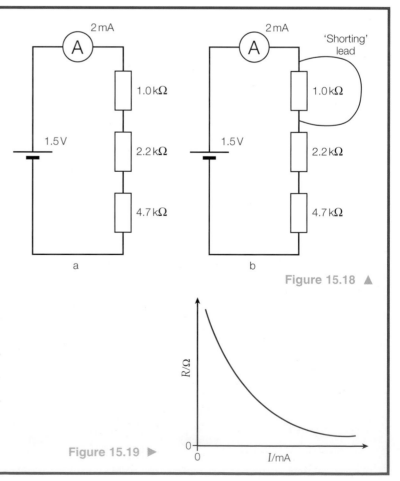

Figure 15.18 ▲

Figure 15.19 ▶

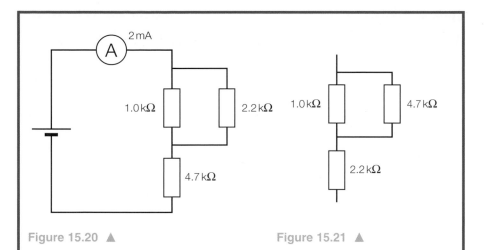

Figure 15.20 ▲   Figure 15.21 ▲

Now set up the circuit shown in Figure 15.20.

Record the current I and use your graph to find a value for the resistance R of this series and parallel combination.

Calculate the theoretical value for R and determine the percentage difference between your experimental value and the calculated value. Comment on your answer in terms of possible sources of experimental error.

A typical set of observations from a similar experiment is shown in Table 15.6.

$R/k\Omega$	7.9	6.9	5.7	4.7	3.3	2.2	1.0	Combination
I/mA	0.19	0.22	0.27	0.33	0.47	0.70	1.54	0.51

Table 15.6 ▲

Plot a graph of these data and use it to show that in this case the resistors were arranged as shown in Figure 15.21 (your experimental and calculated values should both be about $3\,k\Omega$).

15.8 Current and power calculations in series and parallel circuits

Questions involving the calculation of current and power in different series and parallel combinations have to be treated individually. They have to be worked out from first principles using the techniques we have developed in previous examples. There are no short cuts – each problem must be carefully analysed, step-by-step, showing all of your working.

Worked example

In the circuit shown in Figure 15.22 you may assume that the 6.0 V battery has negligible internal resistance.

1 For this circuit, calculate:
 a) the current in each resistor
 b) the power dissipated in each resistor
 c) the power developed by the battery.
 Comment on your answers to parts **1b** and **1c**.

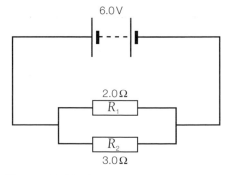

Figure 15.22 ▲

Answer

1 a) The 6.0 V of the battery will be the potential difference across each resistor (as its internal resistance is negligible), so:

$$I_1 = \frac{V}{R_1} = \frac{6.0\,\text{V}}{2.0\,\Omega} = 3.0\,\text{A}$$

$$I_2 = \frac{V}{R_2} = \frac{6.0\,\text{V}}{3.0\,\Omega} = 2.0\,\text{A}$$

b) We can now calculate the power dissipated:

$$P_1 = I_1^2 R_1 = (3.0\,\text{A})^2 \times 2.0\,\Omega = 18\,\text{W}$$
$$P_2 = I_2^2 R_2 = (2.0\,\text{A})^2 \times 3.0\,\Omega = 12\,\text{W}$$

c) The power developed by the battery is given by $P = IV$, where the current in the battery is $I = I_1 + I_2 = 2.0\,\text{A} + 3.0\,\text{A} = 5.0\,\text{A}$, so:

$$P = IV = 5.0\,\text{A} \times 6.0\,\text{V} = 30\,\text{W}$$

In other words, the power developed by the battery = power dissipated in the two resistors. This should come as no surprise – we would expect this to be the case from the law of conservation of energy.

---

2 Repeat your calculations, but this time assume that the battery has an internal resistance of $r = 0.8\,\Omega$.

Answer

2 Redraw the circuit to include the internal resistance of the battery as shown in Figure 15.23.

We have to start by calculating the combined resistance of the parallel arrangement:

$$\frac{1}{R} = \frac{1}{2.0\,\Omega} + \frac{1}{3.0\,\Omega} = 0.5\,\Omega^{-1} + 0.33\,\Omega^{-1} = 0.83\,\Omega^{-1}$$

$$R = \frac{1}{0.83\,\Omega^{-1}} = 1.2\,\Omega$$

The total circuit resistance R_T is now calculated as:

$$R_T = 1.2\,\Omega + 0.8\,\Omega = 2.0\,\Omega$$

We can now calculate the circuit current I from:

$$I = \frac{\varepsilon}{R} = \frac{6.0\,\text{V}}{2.0\,\Omega} = 3.0\,\text{A}$$

The potential difference across the parallel resistors is given by:

$$V = \varepsilon - Ir = 6.0\,\text{V} - (3.0\,\text{A} \times 0.8\,\Omega) = 6.0\,\text{V} - 2.4\,\text{V} = 3.6\,\text{V}$$

Hence:

$$I_1 = \frac{V}{R_1} = \frac{3.6\,\text{V}}{2.0\,\Omega} = 1.8\,\text{A} \qquad I_2 = \frac{V}{R_2} = \frac{3.6\,\text{V}}{3.0\,\Omega} = 1.2\,\text{A}$$

$$P_1 = I_1^2 R_1 = (1.8\,\text{A})^2 \times 2.0\,\Omega = 6.5\,\text{W}$$
$$P_2 = I_2^2 R_2 = (1.2\,\text{A})^2 \times 3.0\,\Omega = 4.3\,\text{W}$$

The power developed by the battery is $P = \varepsilon I = 6.0\,\text{V} \times 3.0\,\text{A} = 18.0\,\text{W}$, while the power dissipated in the two resistors is only $6.5\,\text{W} + 4.3\,\text{W} = 10.8\,\text{W}$. What has happened to the other 7.2 W?

The other 7.2 W is the power 'wasted' in the battery – the work done per second by the charge overcoming the internal resistance as it passes through the battery:

$$\text{'wasted' power} = I^2 r = (3.0\,\text{A})^2 \times 0.8\,\Omega = 7.2\,\text{W}$$

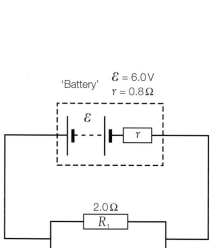

'Battery' $\varepsilon = 6.0\,\text{V}$
$r = 0.8\,\Omega$

ε

r

$2.0\,\Omega$
R_1

R_2
$3.0\,\Omega$

Figure 15.23 ▲

15.9 Principle of the potential divider

We can see how a potential divider works by considering the circuit shown in Figure 15.24.

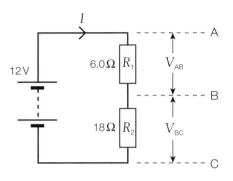

Figure 15.24 ▲

$$\text{Current, } I = \frac{V}{R} = \frac{12\,V}{24\,\Omega} = 0.50\,A$$

$$V_{AB} = IR_1 = 0.50\,A \times 6.0\,\Omega = 3.0\,V$$

$$V_{BC} = IR_2 = 0.50\,A \times 18\,\Omega = 9.0\,V$$

The network of resistors R_1 and R_2 has **divided** the potential difference of 12.0 V into 3.0 V and 9.0 V across R_1 and R_2, respectively. Such an arrangement is called a **potential divider**.

If you are mathematically minded, and good at proportions, you will see that the potential difference has been divided in the **ratio** of the resistances (1:3). This is a quick and easy way of calculating the potential differences in such an arrangement, but only if the ratio is a simple one!

Worked example

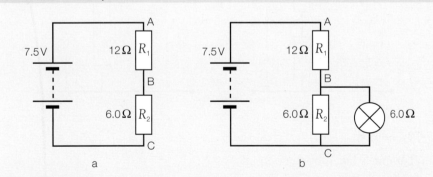

Figure 15.25 ▲

In the circuits shown in Figure 15.25, you may assume that the battery has negligible internal resistance.

1 Calculate the potential difference across the 6.0 Ω resistor in the circuit shown in Figure 15.25a.

2 A lamp of resistance 6.0 Ω is now connected across the 6.0 Ω resistor as shown in Figure 15.25b. What is:

   a) the p.d. across the lamp

   b) the current in the lamp?

Answer

1 Circuit current:

$$\text{Current, } I = \frac{V}{R} = \frac{7.5\,V}{18\,\Omega} = 0.4166...\,A$$

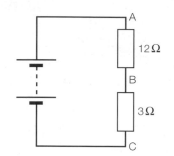

Figure 15.26 ▲

Across the $6\,\Omega$ resistor:

$$V = IR = 0.4166\ldots\,A \times 6.0\,\Omega = 2.5\,V$$

2 a) It would be tempting (but wrong!) to say that this p.d. of 2.5 V across R_2 would also be the p.d. across the lamp in Figure 15.25b. However, the inclusion of the lamp changes the resistance between BC:

$$\frac{1}{R} = \frac{1}{6.0\,\Omega} + \frac{1}{6.0\,\Omega} = \frac{2}{6.0\,\Omega}$$

$$R_{BC} = 3.0\,\Omega$$

The equivalent circuit is now as shown in Figure 15.26.

The circuit current becomes:

$$I = \frac{V}{R} = \frac{7.5\,V}{15\,\Omega} = 0.5\,A$$

$$V_{BC} = IR_{BC} = 0.50\,A \times 3.0\,\Omega = 1.5\,V$$

b) The current in the lamp is:

$$I = \frac{V}{R} = \frac{1.5\,V}{6\,\Omega} = 0.25\,A$$

This worked example shows how adding a 'load' to the output can affect the output voltage. In order to minimise this effect, a general rule of thumb is that the resistance of the 'load' should be at least **10 times** greater than the output resistance (R_2 in the example).

We can now formulate a general expression for a potential divider circuit by considering Figure 15.27a.

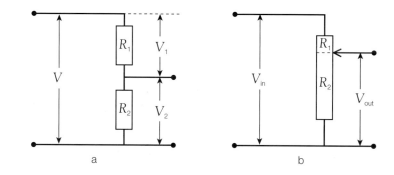

Figure 15.27 ▲

If no current is taken by the output:

$$I = \frac{V_2}{R_2} \text{ and also } I = \frac{V}{R_1 + R_2}$$

$$\Rightarrow \frac{V_2}{R_2} = \frac{V}{R_1 + R_2}$$

so:

$$V_2 = V \times \frac{R_2}{R_1 + R_2} \text{ and similarly } V_1 = V \times \frac{R_1}{R_1 + R_2}$$

In Figure 15.27b, the two resistors have been replaced by a three-terminal variable resistor. We can think of the sliding contact as dividing this resistor into two imaginary resistances R_1 and R_2. As the sliding contact is moved up, 'R_2' gets larger and 'R_1' gets smaller (but of course $R_1 + R_2$ remains the same).

As $V_{out} = V_{in} \times \dfrac{R_2}{R_1 + R_2}$

Tip

Although remembering these formulae allows you to work out simple examples very easily and quickly, it is recommended that you also learn to tackle problems from first principles – as shown in the worked example.

V_{out} will get progressively larger as the slider is moved up. With the slider right at the top, $V_{out} = V_{in}$; when the slider is right at the bottom, $V_{out} = 0$. Thus a continuously variable range of output voltages from 0 to V_{in} can be obtained. You can check this for yourself by setting up the circuit shown in Figure 15.27b. A 1.5 V cell can be used for V_{in}, and a digital voltmeter can be used to measure V_{out}.

15.10 Practical use of a potential divider

We saw how a **rheostat** (Figure 15.28a) could be used to provide a continuously variable potential difference in Section 13.1. A three-terminal resistor used as a potential divider is called a **potentiometer**. A potentiometer such as that shown in Figure 15.28b is, in effect, a rheostat 'bent' around to form an almost complete circle. It has a fixed contact at each end and a rotating arm that forms the sliding contact.

Although a 'rheostat' is often used in school laboratories as a potential divider, it should really be called a 'potentiometer' when used in this way. Commercial potentiometers can be circular, as shown in Figure 15.28b, or straight – for example radio volume controls can either be rotating knobs or sliders.

We also saw that a rheostat could be used either to vary current (Figure 15.29a) or to vary potential difference (Figure 15.29b):

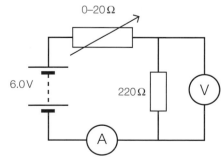

Figure 15.28 ▲
a) A rheostat and b) a potentiometer

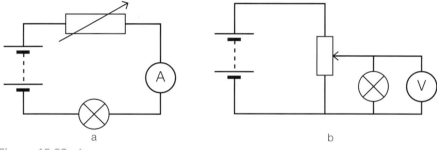

Figure 15.29 ▲

We showed experimentally that the circuit in Figure 15.29b was the more useful as we could get a much greater variation in the potential difference – from zero up to the p.d. of the battery. To refresh your memory, look at the following worked example.

Worked example

A student proposes to use the circuit shown in Figure 15.30 to investigate the *I–V* characteristics of a $220\,\Omega$ resistor.

1 To decide whether the circuit is appropriate, determine:

   a) the maximum voltage that can be obtained across the $220\,\Omega$ resistor

   b) the minimum voltage that can be obtained across the $220\,\Omega$ resistor.

Answer

1 **a)** When the rheostat is set to its minimum resistance ($0\,\Omega$), the voltage across the $220\,\Omega$ resistor will have a maximum value equal to the p.d. of the battery – that is, 6.0 V.

   b) When the rheostat is set to its maximum resistance of $20\,\Omega$, the circuit resistance will have a maximum value of:

$$R = 20\,\Omega + 220\,\Omega = 240\,\Omega$$

   which will give a minimum circuit current of:

$$I = \frac{V}{R} = \frac{6.0\,\text{V}}{240\,\Omega} = 0.025\,\text{A}$$

Figure 15.30 ▲

and a minimum voltage across the $220\,\Omega$ resistor of:

$V_{min} = IR = 0.025\,\text{A} \times 220\,\Omega = 5.5\,\text{V}$

The student could therefore only investigate the I–V characteristics over a voltage range of 5.5–6.0 V – clearly not very satisfactory!

2 The student's teacher suggests that the same apparatus could be rearranged to form a potential divider circuit, which would enable a voltage range of between 0 V and 6.0 V to be achieved. Draw a circuit diagram to show how this could be done.

Answer

2

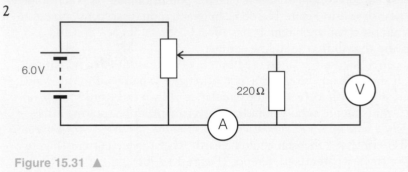

Figure 15.31 ▲

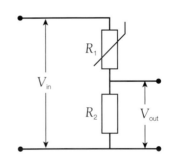

Figure 15.32 ▲

15.11 Using a thermistor to control voltage

A thermistor can be used in a potential divider circuit to control the output voltage as shown in Figure 15.32.

From the formula derived in Section 15.9:

$$V_{out} = V_{in} \times \frac{R_2}{R_1 + R_2}$$

if the temperature rises, the resistance R_1 of the thermistor will decrease and so V_{out} will increase.

Looking at the situation from first principles, if R_1 decreases, the circuit resistance will get less and so the circuit current, I, will get larger. As $V_{out} = IR_2$, V_{out} will also increase.

Worked example

The thermistor in Figure 15.33 has a resistance of:

1 $440\,\Omega$ at 20 °C

2 $110\,\Omega$ at 60 °C

Calculate the value of the output voltage at each of these resistances.

Answer

1 At 20 °C, where $R_1 = 440\,\Omega$:

$$V_{out} = V_{in} \times \frac{R_2}{R_1 + R_2}$$

$$= 6.0\,\text{V} \times \frac{220\,\Omega}{440\,\Omega + 220\,\Omega} = 6.0\,\text{V} \times \frac{1}{3} = 2.0\,\text{V}$$

2 At 60 °C, where $R_1 = 110\,\Omega$:

$$V_{out} = V_{in} \times \frac{R_2}{R_1 + R_2}$$

$$= 6.0\,\text{V} \times \frac{220\,\Omega}{110\,\Omega + 220\,\Omega} = 6.0\,\text{V} \times \frac{2}{3} = 4.0\,\text{V}$$

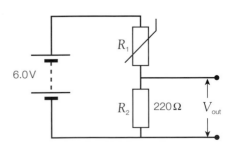

Figure 15.33 ▲

Experiment

Using a thermistor in a potential divider circuit to activate a thermostatic switch

Use an ohmmeter to measure the resistance R_1 of the thermistor at room temperature. Now set up the arrangement shown in Figure 15.34a; the circuit diagram for this arrangement is shown in Figure 15.34b. The resistor should be selected such that its resistance $R_2 \approx \frac{1}{2} R_1$.

Measure the output voltage V_{out} for temperatures θ ranging from room temperature to about 60 °C. Plot a graph of V_{out} against θ and use it to determine the value of V_{out} when the temperature is 40 °C.

Figure 15.34 ▶

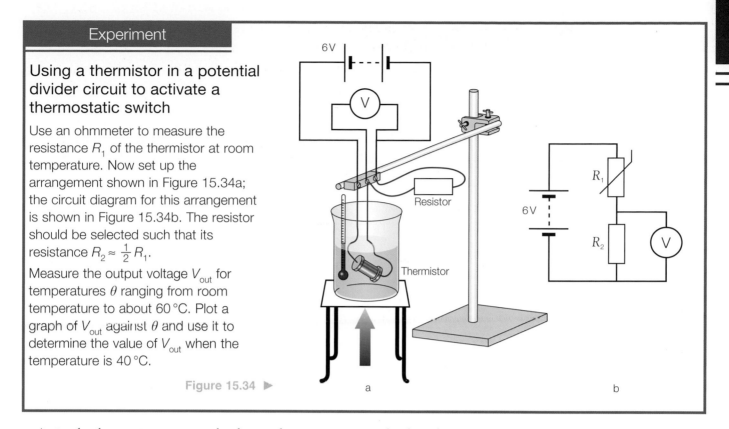

a

b

A simple electronic circuit can be designed to operate a switch when the output from the circuit in Figure 15.34b reaches a predetermined value. For example, if this were to be the value of output voltage you obtained in your experiment, such a circuit could be used to switch on a warning lamp when the temperature reached 40 °C.

REVIEW QUESTIONS

1 Figure 15.35 shows four possible ways of connecting three $15\,\Omega$ resistors.

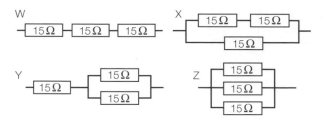

Figure 15.35 ▲

a) The resistance of combination Z is:

A $\dfrac{W}{9}$ B $\dfrac{W}{3}$

C $3\,W$ D $9\,W$

b) The resistance of combination W is:

A $\dfrac{X}{2}$ B $2X$

C $\dfrac{Y}{9}$ D $2Y$

c) The resistance of combination X is:

A $\frac{1}{2} \times Y$ B $2 \times Y$

C $\frac{1}{2} \times Z$ D $2 \times Z$

2 a) Calculate the resistance of the combination of resistors shown in Figure 15.36.

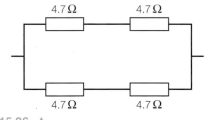

Figure 15.36 ▲

b) Comment on your answer and suggest why this network might be used rather than a single resistor of the same value.

3 A student has a 9 V battery in his pocket with some coins. The battery is accidentally 'short-circuited' by a coin. Calculate the current that will flow in the battery if it has an internal resistance of $0.50\,\Omega$ and hence explain why the battery gets hot.

4 A laboratory high-voltage supply has a $50\,M\Omega$ resistor inside it to effectively give it a large internal resistance.

a) Explain why it has this resistor.

b) What is the maximum current that the supply can give when the voltage is set to 5 kV?

c) A girl has a resistance of $10\,k\Omega$, as measured between her hand and the ground. Explain why, if she were to accidentally touch the live terminal of the high-voltage supply, the current in her would be virtually the same as in part b).

d) Calculate the power that would be dissipated in the girl. Comment on this value compared with the 14 mW from the car battery in the example on page 156.

5 Copy the table of data from the experiment on page 155 and add columns for the resistance $R = \dfrac{V}{I}$ of the variable resistor and the corresponding power $P = IV$ dissipated in it. Then plot a graph of P against R. You should find that the power has a maximum value when the resistance of the variable resistor (the 'load') is equal to the internal resistance of the cell (the 'power supply'). This is always the case and is important in electronics for 'matching' the load (such as loudspeakers) to the power supply (such as an amplifier).

6 The data in Table 15.7 were obtained for the output of a solar cell using the circuit shown in Figure 15.7 when the cell was illuminated by a bench lamp.

I/mA	0.10	0.15	0.20	0.30	0.40	0.50	0.60	0.65
V/V	6.07	5.90	5.72	5.35	4.97	4.20	3.00	2.01
R/kΩ								
P/mW								

Table 15.7 ▲

a) Complete Table 15.7 by adding values for the resistance across the cell, $R = \dfrac{V}{I}$, and the power generated, $P = VI$.

b) i) Draw a graph of V against I and comment on the shape of your graph.

ii) Use the graph to determine values for the e.m.f. of the cell and the internal resistance of the cell for low current values.

c) i) Plot a graph of P against R.

ii) Use the graph to determine the maximum power generated.

7 A student proposes to set up the circuit shown in Figure 15.37. The battery may be assumed to have negligible internal resistance.

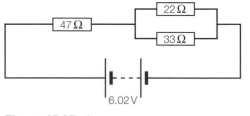

Figure 15.37 ▲

a) What would be:

   i) the current in each resistor

   ii) the power generated in each resistor?

b) Discuss whether resistors with a power rating of 500 mW would be suitable for the student to use in this circuit.

8 A student sets up the circuit shown in Figure 15.38 with a 10 V analogue voltmeter:

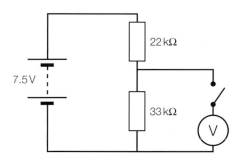

Figure 15.38 ▲

a) Show that the potential difference across the 33 kΩ resistor before the voltmeter is connected is 4.5 V.

b) When the switch is closed, the voltmeter reads 4.0 V.

   i) Account for the difference between this value and your answer to part **a)**.

   ii) Calculate the resistance of the voltmeter.

c) The voltmeter scale is marked 10 V/100 μA and the resistors have a 5% tolerance. Discuss whether your answer to part **b)ii)** is compatible with this data.

9 Figure 15.39a shows a circuit containing a thermistor. Figure 15.39b shows how the resistance of the thermistor varies with temperature.

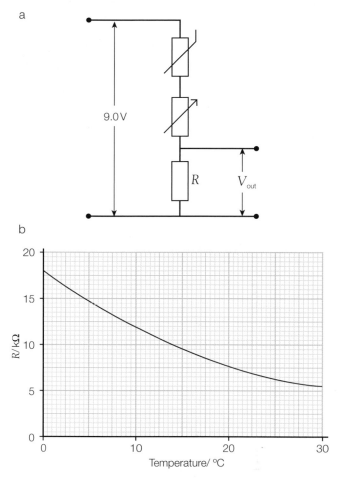

Figure 15.39 ▲

a) Explain why V_{out} gets less as the temperature of the thermistor falls.

b) The circuit is to be used to switch on a warning lamp when the temperature drops to 0 °C. The switch is activated when V_{out} is 5.0 V.

   i) Calculate the value of the resistor R that you would use given that preferred values (that is, standard values supplied by manufacturers) of 10 kΩ, 22 kΩ, 33 kΩ and 47 kΩ are available.

   ii) Explain the purpose of the potentiometer.

16 Nature of light

Figure 16.1 ▲
a) Sir Isaac Newton and b) Christiaan Huygens

16.1 Some early theories

'And God said, "Let there be light"; and there was light.' These words appear right at the very beginning of the Bible and reflect the importance that light has in all religions. During Diwali – the 'festival of lights' – Hindus celebrate the victory of good over evil and of light over darkness. The Koran says that 'God is the light of Heaven and Earth'. Evidence that very early religions worshipped light is provided by great monuments such as the Pyramids and Stonehenge.

Without light we would not even exist. Life depends on three things: long-chain carbon molecules, water and light. The Earth had all three, and so within the oceans a rich organic soup that ultimately bore life began to form. It is no wonder that scientists have been fascinated by light, and many have devoted much of their lives in trying to unravel its mysteries.

Some 2500 years ago, the Greek philosophers were divided in their opinions as to the nature of light. Pythagoras suggested that visible objects emitted a stream of particles that bombarded the eye, while Plato reasoned that light originated from the eye and then reflected off objects to enable us to see them. Euclid first put forward the concept of light as rays travelling in straight lines, while Aristotle thought of light as waves and tried to relate colour to music (as did Newton some 2000 years later).

When Western science, like art and literature, emerged from the dark ages (no pun intended!), the scientists of the day once more debated the nature of light. In the seventeenth century, great men such as Galileo, Kepler, Huygens and Newton pushed back the barriers of physics and astronomy with their experiments and theories. Galileo attempted, unsuccessfully, to measure the speed of light and concluded that if it was not infinite (as was thought to be the case by many), it was certainly very fast. It was left to the Danish astronomer Olaus Roemer to make the first determination of the speed of light in 1676, 34 years after Galileo's death. His value of 200 million $m\,s^{-1}$ was remarkably close to today's defined value of $299\,792\,458\,m\,s^{-1}$. Yes, we **define** the speed of light in a vacuum to nine significant figures as it is such a fundamental constant within our universe! Since 1983, the **metre** has been defined as being the distance travelled by light in a vacuum in $1/299\,792\,458\,s$. This means that the speed of light c is now a defined quantity and distances are measured accurately using laser pulses and atomic clocks.

Meanwhile, the nature of light was being fiercely debated by Newton and Huygens. Huygens, like Newton, also contributed much to the development of mechanics and calculus. Newton proposed the **corpuscular** theory, in which he thought of light as being made up of corpuscles (or tiny particles). Using this theory, he was able to give a mathematical interpretation of the laws of reflection and refraction. He imagined the light particles to bounce off surfaces, like a ball bouncing off a wall, to explain reflection and suggested that the light particles changed speed when they moved from one material to another to explain refraction.

Huygens put forward a completely different idea, however. He maintained that light consisted of waves, like ripples spreading out when a stone is dropped in a pond. His **wave theory** was also able to explain the laws of reflection and refraction, with one fundamental difference from Newton – the wave theory required the speed of light to be **less** in glass or water than in air, whereas in the particle theory the opposite was the case.

Further evidence for the wave theory was provided by Thomas Young around 1800. His famous two-slit experiment demonstrated the phenomenon of interference (see Chapter 10), which could be satisfactorily explained at the time only by considering light to be in the form of waves. There was only one way to resolve the argument – to measure the speed of light in air and then in glass or water!

That's just what Foucault did. In 1850, he showed qualitatively that the speed of light in water was indeed slower than it was in air. Further, more accurate, measurements confirmed that the ratio of the speed of light in air to that in water was equal to the refractive index of water – exactly as predicted by the wave theory.

About the same time, James Clerk Maxwell developed a model for an electromagnetic wave based on the mathematics of electric and magnetic fields (which are considered in the A-level course). From Maxwell's equation for an electromagnetic wave, it was possible to calculate the speed of the wave from the electric and magnetic field constants. This gave a value for the speed of light that agreed exactly with that found experimentally.

The wave theory proposed the existence of a fluid medium, called the aether, necessary for transmission of the waves. A celebrated experiment performed by Michelson and Morley in 1887 failed to detect the motion of the Earth through the aether, and it was left to the genius of Albert Einstein to resolve the matter. In 1905, he took the Michelson–Morley experiment as the starting point of his theory of relativity. This theory, which concluded that all motion was relative, denied the existence of any stationary medium and so effectively abolished the aether. Game, set and match to the wave theory…or was it?

16.2 Intensity of light

In everyday usage, the word 'intensity' means the strength or level of something. In physics, it has a more precise meaning. We define:

$$\text{intensity} = \frac{\text{power}}{\text{area}}$$

so intensity has units of $W\,m^{-2}$.

When we talk about the radiation from the Sun falling on the Earth, we sometimes use the phrase **radiation flux density** instead of intensity. You need to be familiar with both terms.

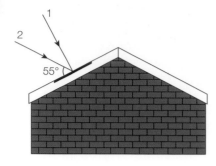

Figure 16.2 ▲

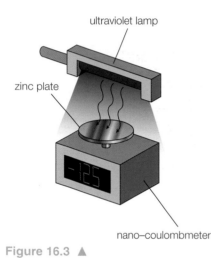

Figure 16.3 ▲

Worked example

A solar heating panel has an area of $2.8\,m^2$. How much energy falls on the panel per second if the solar radiation flux density is $300\,W\,m^{-2}$ when:

1 the panel is normal to the Sun's radiation

2 the radiation makes an angle of $55°$ with the panel?

Answer

1 Power = radiation flux density × area = $300\,W\,m^{-2} \times 2.8\,m^2 = 840\,J\,s^{-1}$

2 If the angle between the radiation and the panel is $55°$, we need to consider the component of the radiation that falls at right angles (that is, 'normal') to the panel, so the energy reduces to:

$840 \sin 55° = 690\,J\,s^{-1}$

If we are a distance r from a point source of radiation, the radiation will have spread out in all directions over a sphere of surface area $4\pi r^2$ by the time it reaches us. The intensity will therefore be:

$$\text{intensity} = \frac{\text{power}}{4\pi r^2}$$

Worked example

The solar constant (the power of the radiation from the Sun falling normally on $1\,m^2$ of the outer surface of the Earth's atmosphere) is approximately $1.4\,kW\,m^{-2}$. Estimate the power radiated by the Sun assuming that it is 150 million km away.

Answer

$$\text{intensity} = \frac{\text{power}}{4\pi r^2} \Rightarrow \text{power} = \text{intensity} \times 4\pi r^2$$
$$= 1.4 \times 10^3\,W\,m^{-2} \times 4\pi \times (150 \times 10^6 \times 10^3\,m)^2$$
$$= 4.0 \times 10^{26}\,W$$

16.3 Photoelectric effect

The photoelectric effect can be demonstrated by means of a negatively charged zinc plate, as shown in Figure 16.3.

A freshly cleaned zinc plate is given a negative charge (for example, with a polythene rod – see Figure 11.1). The nano-coulombmeter indicates that in visible light, even from a 60 W lamp shining directly onto the zinc plate, the zinc does not discharge (or at least only very slowly if the atmosphere is humid). However, if even a weak ultraviolet light is shone onto the zinc plate, it begins to discharge immediately.

This phenomenon is called the photoelectric effect – light ('photo') causing the emission of electrons ('electric'). Energy from the light is given to the electrons in the zinc, and some electrons near the surface of the zinc gain enough energy to escape from the attraction of the positive charge on the nucleus. These electrons are called photoelectrons. As the visible light is much more intense (and therefore has more energy) than the ultraviolet light, how can we explain why this happens with the ultraviolet and not the visible light?

Once more it is Einstein to the rescue! In 1905, he explained the photoelectric effect in terms of a **quantum theory**. The word 'quantum' means a fixed indivisible amount. For example, we say that charge is quantised because all electric charges are made up of a whole number multiple of 'fixed' electron charges (that is, $1.60 \times 10^{-19}\,C$). Five years earlier, Max Planck had proposed mathematically that electromagnetic radiation did not exist as continuous waves but as discrete bundles (quanta) of energy, which we now call **photons**. Each photon has a discrete amount of energy hf, where f is the frequency of the radiation and h is a universal constant, now called the Planck constant ($h = 6.63 \times 10^{-34}$ J s). This theory enabled Planck to successfully explain the distribution of energy with wavelength for the radiation from a hot body. Einstein applied Planck's theory to the photoelectric effect.

Instead of the energy from the light being gradually absorbed by the electrons near the surface of the zinc until they had enough energy to escape, Einstein reasoned that the electron would be emitted only if a **single quantum** of the light had enough energy for the electron to escape. The frequency of visible light is not high enough for a photon (energy $= hf$) to provide the necessary energy for an electron to escape; however, ultraviolet light has a higher frequency than visible light. Its quanta each have sufficient energy to release an electron, so the zinc plate begins to discharge immediately in the presence of ultraviolet light.

The **intensity** of a light source depends on both the number of quanta and the energy associated with each quantum. However intense a visible light source may be, the energy of each of its quanta will be insufficient to liberate an electron from the zinc, so the zinc plate will not discharge in the presence of visible light.

> **Tip**
>
> Remember that **one** photon (if it has enough energy) will release **one** photoelectron.

Worked example

An advertisement for a green laser pointer gives the data in Table 16.1.

1 What is the efficiency of the laser in converting electrical energy to light energy?

2 What is the intensity (radiation flux density) of the light beam close to the laser?

3 How much energy is in one photon of the light?

4 How many photons per second are emitted by the laser?

5 The manufacturer claims that this laser is 'much brighter than a 5 mW red laser'. Suggest why this statement may be justified.

Wavelength	532 nm
Power	5 mW
Beam diameter at source	1.5 mm
Current	300 mA
Power supply	2 × 1.5 V AAA cells

Table 16.1 ▲

Answer

1 Electrical power consumed $= VI = 3.0$ V $\times 300 \times 10^{-3}$ A $= 0.9$ W (or 900 mW)

$$\text{Efficiency} = \frac{\text{output}}{\text{input}} \times 100\% = \frac{5\,\text{mW}}{900\,\text{mW}} \times 100\% = 0.5\%$$

2 $$\text{Intensity} = \frac{\text{power}}{\text{area}} = \frac{5 \times 10^{-3}\,\text{W}}{\pi(0.5 \times 1.5 \times 10^{-3}\,\text{m})^2} = 2.8\,\text{kW m}^{-2}$$

3 Energy of 1 photon, $E = hf$, where f is given by $c = f\lambda$

$$f = \frac{c}{\lambda} = \frac{3.00 \times 10^8\,\text{m s}^{-1}}{532 \times 10^{-9}\,\text{m}} = 5.64 \times 10^{14}\,\text{Hz}$$

$$E = hf = 6.63 \times 10^{-34}\,\text{J s} \times 5.64 \times 10^{14}\,\text{s}^{-1} = 3.7 \times 10^{-19}\,\text{J}$$

4 Energy emitted per second $= 5$ mW $= 5 \times 10^{-3}$ J s^{-1}

$$\text{Number of photons emitted per second} = \frac{5 \times 10^{-3}\,\text{J s}^{-1}}{3.7 \times 10^{-19}\,\text{J}} = 1.4 \times 10^{16}\,\text{s}^{-1}$$

5 Green light (wavelength 532 nm) is in the middle of the visible spectrum (400–700 nm). As the human eye is most sensitive to light in this region, the green laser will **seem** much brighter than a red laser of comparable power, so there may be some justification in the manufacturer's claim.

Tip

Remember:
$1\,\mathrm{eV} \equiv 1.6 \times 10^{-19}\,\mathrm{J}$

16.4 Electron-volt

In the above calculation, we found that the energy of a photon of green light was $3.7 \times 10^{-19}\,\mathrm{J}$. This is a very small amount of energy, so rather than keep having to include powers of 10^{-19} when considering photon energy, we often use the electron-volt (eV) as a convenient unit of energy. An electron-volt is the work done on (or the energy gained by) an electron when it moves through a potential difference of 1 volt. From $W = QV$:

$$1 \text{ electron-volt} = 1.6 \times 10^{-19}\,\mathrm{C} \times 1\,\mathrm{V} = 1.6 \times 10^{-19}\,\mathrm{J}$$

As the electron-volt is a very small unit of energy, we often use keV ($10^3\,\mathrm{eV}$) and MeV ($10^6\,\mathrm{eV}$). For example, typical X-rays have energy of 120 keV and alpha particles from americium-241 (commonly found in smoke detectors) have energy of 5.6 MeV.

Worked example

Calculate:

1 the energy, in electron-volts, of a photon of light from a red laser pointer with wavelength 650 nm

2 the wavelength of a 120 keV X-ray.

Answer

1 $E = hf = \dfrac{hc}{\lambda} = \dfrac{6.63 \times 10^{-34}\,\mathrm{J\,s} \times 3.00 \times 10^{8}\,\mathrm{m\,s^{-1}}}{650 \times 10^{-9}\,\mathrm{m}}$

$\quad = 3.06 \times 10^{-19}\,\mathrm{J}$

$\quad = \dfrac{3.06 \times 10^{-19}\,\mathrm{J}}{1.6 \times 10^{-19}\,\mathrm{J\,eV^{-1}}} = 1.9\,\mathrm{eV}$

2 $E = 120\,\mathrm{keV} = 120 \times 10^{3}\,\mathrm{eV} \times 1.6 \times 10^{-19}\,\mathrm{J\,eV^{-1}} = 1.92 \times 10^{-14}\,\mathrm{J}$

$\quad E = hf = \dfrac{hc}{\lambda}$

$\Rightarrow \lambda = \dfrac{hc}{E} = \dfrac{6.63 \times 10^{-34}\,\mathrm{J\,s} \times 3.00 \times 10^{8}\,\mathrm{m\,s^{-1}}}{1.92 \times 10^{-14}\,\mathrm{J}}$

$\quad = 1.0 \times 10^{-11}\,\mathrm{m}$

16.5 Einstein's photoelectric equation

If a photon of energy hf has more than the bare minimum energy needed to just remove an electron from the surface of a metal (called the **work function**, symbol ϕ), the remaining energy is given to the electron as kinetic energy, $\frac{1}{2}mv^2$.

Applying the conservation of energy, the maximum kinetic energy $\frac{1}{2}mv_{max}^2$ that an electron can have is given by:

$$hf = \phi + \frac{1}{2}mv_{max}^2$$

This is known as Einstein's photoelectric equation. Electrons emitted from further inside the metal will need more than the work function to escape and so will have less than this maximum kinetic energy. Therefore, electrons are emitted with a range of kinetic energies up to the maximum defined by the equation.

Einstein's equation indicates that there will be no photoelectric emission unless $hf > \phi$. The frequency that is just large enough to liberate electrons, f_0, is called the **threshold frequency**, so $\phi = hf_0$.

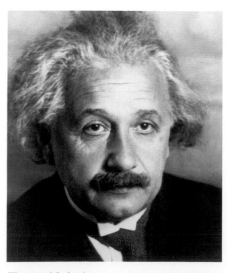

Figure 16.4 ▲
Albert Einstein

Worked example

The work function for zinc is 4.3 eV. Explain why photoelectric emission is observed when ultraviolet light of wavelength in the order of 200 nm is shone onto a zinc plate but not when a 60 W filament lamp is used.

Answer
A photon of the ultraviolet light has energy:

$$E = hf = \frac{hc}{\lambda} = \frac{6.63 \times 10^{-34}\,\text{J s} \times 3.00 \times 10^8\,\text{m s}^{-1}}{200 \times 10^{-9}\,\text{m}} = 9.95 \times 10^{-19}\,\text{J}$$

$$= \frac{9.95 \times 10^{-19}\,\text{J}}{1.6 \times 10^{-19}\,\text{J eV}^{-1}} = 6.2\,\text{eV}$$

This is greater than the 4.3 eV work function for zinc. This means that each photon will have sufficient energy to remove an electron from the surface of the zinc and photoemission will occur.

The shortest wavelength (highest frequency) visible light is 400 nm (at the blue end of the visible spectrum). The threshold frequency f_0 for zinc is given by:

$$\psi = hf_0 \implies f_0 = \frac{\phi}{h} = \frac{4.3\,\text{eV} \times 1.6 \times 10^{-19}\,\text{J eV}^{-1}}{6.63 \times 10^{-34}\,\text{J s}} = 1.0 \times 10^{15}\,\text{Hz}$$

The frequency of the blue end of the visible spectrum is:

$$f = \frac{c}{\lambda} = \frac{3.00 \times 10^8\,\text{m s}^{-1}}{400 \times 10^{-9}\,\text{m}} = 7.5 \times 10^{14}\,\text{Hz} \text{ (that is, } <1.0 \times 10^{15}\,\text{Hz)}$$

Photoemission therefore will not take place.

16.6 Phototube

A phototube is the name given to a particular type of **photocell** that generates photoelectrons when light falls on a specially coated metal cathode. The other types of photocells are **photovoltaic** photocells, in which an e.m.f. is generated by the presence of light across the boundary of two semiconducting materials, and **photoconductive** cells, or **light-dependent resistors** (LDRs). An LDR is a semiconductor whose resistance decreases (that is, it becomes a better conductor) when it is exposed to electromagnetic radiation.

Extension

A phototube can be used to investigate the photoelectric effect by connecting a variable potential difference across it and measuring the current in it with a very sensitive ammeter, as shown in Figure 16.5.

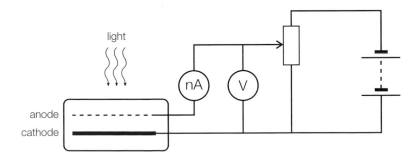

Figure 16.5 ▲

If the power supply is connected so that the anode is **negative**, the reverse potential difference does work slowing down the photoelectrons. When the work done is equal to the maximum kinetic energy of the photoelectrons, they will not reach the anode and the current will be zero. The reverse potential to just do this is called the **stopping potential**, V_s. It is given by:

$$eV_s = \frac{1}{2}mv_{max}^2$$

Measuring the stopping potential enables us to determine the maximum kinetic energy of the photoelectrons.

Worked example

The graph in Figure 16.6 shows how the maximum kinetic energy $\frac{1}{2}mv_{max}^2$ of photoelectrons emitted from the surface of caesium varies with the frequency, f, of the incident electromagnetic radiation.

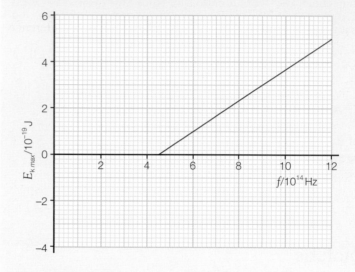

Figure 16.6 ▲

Use the graph to find:

1 the value for the Planck constant

2 the work function ϕ for caesium.

Answer

1 Rearranging $hf = \phi + \frac{1}{2}mv_{max}^2$ gives:

$$\frac{1}{2}mv_{max} = hf - \phi$$

Comparing $\quad y = mx + c$

h will be the gradient of the graph and $-\phi$ will be the intercept:

$$h = \frac{(5.0 - 0.0) \times 10^{-19}\,\text{J}}{(12.0 - 4.5) \times 10^{-14}\,\text{s}^{-1}} = 6.7 \times 10^{-34}\,\text{J s}$$

2 Extrapolating the graph to the y axis gives $-\phi = -3.0 \times 10^{-19}\,\text{J}$

$$\phi = \frac{3.0 \times 10^{-19}\,\text{J}}{1.6 \times 10^{-19}\,\text{J eV}^{-1}} = 1.9\,\text{eV}$$

16.7 Atomic spectra

A flame can be coloured by holding the salts of certain metals in it: a yellow flame is produced by sodium, a lilac flame by potassium and a green flame by barium. If the flame is viewed through a diffraction grating, a series of bright lines is observed. Sodium gives two yellow lines close together, potassium a red line and a violet line and barium several lines running from red to violet – the brightest being green. Such spectra are called **emission spectra** as light is being emitted, or given out.

In the middle of the nineteenth century, Kirchhoff established that the spectra were characteristic of the atoms and molecules that produce them. In his researches Kirchhoff collaborated with the chemist Bunsen, who provided him with chemicals of the highest purity and the burner that now bears his name, in order to provide a suitable flame.

Spectral lines can also be produced by applying a large potential difference between the ends of a tube filled with gas at low pressure. Such devices are called gas discharge tubes, and they can be seen all around us in the familiar form of yellow street lights. These appear yellow as they contain sodium, which predominantly emits deep yellow wavelengths at 589.0 nm and 589.6 nm. Such light is considered safer to drive under than white light, as its monochromatic output improves the perception of contrast and allows the light to penetrate fog and rain with the minimum of dispersion.

When a hydrogen-filled tube is observed through a diffraction grating, four lines can be identified: two blue lines, a bluish-green line and a red line. Balmer analysed the frequency of these lines and discovered in 1885 that the frequency f of the hydrogen lines was given by a simple mathematical formula. Balmer's equation was modified by Rydberg, who wrote it in the form:

$$f = R \left(\frac{1}{2^2} - \frac{1}{m^2} \right)$$

where R is a constant (now called the Rydberg constant) and m had the values 3, 4, 5 and 6 for each of the lines in turn.

Figure 16.7a shows the visible spectrum. Figure 16.7b shows the **emission** spectrum of hydrogen and Figure 16.7c shows the **absorption** spectrum for hydrogen. This is discussed on page 182.

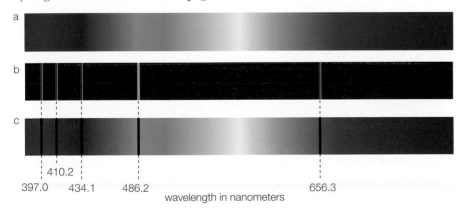

Figure 16.7 ▲ Hydrogen spectrum

Balmer performed some elegant mathematical analysis without the aid of a calculator, let alone a computer, but he was unable to explain why the frequencies were given by a mathematical series formed by the difference of two discrete terms. The mystery was solved by Bohr in 1913. He gave a simple and remarkably exact explanation of the hydrogen spectrum by applying quantum theory to the model of the nuclear atom proposed two years earlier by Rutherford, according to which the hydrogen atom consists of a single electron revolving around a proton. The Rutherford model is discussed in more detail in Unit 4 of the A2 course.

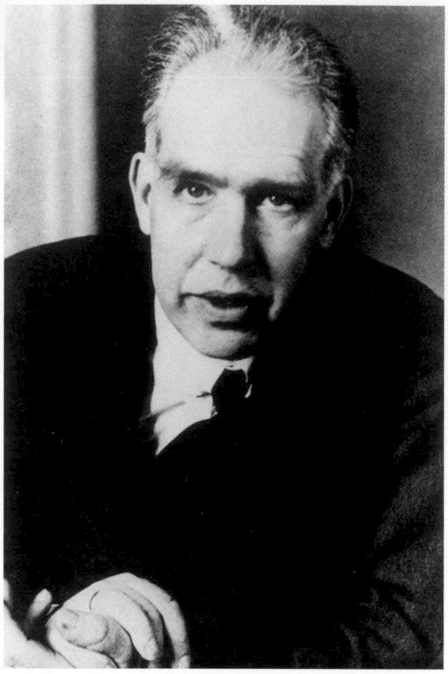

Figure 16.8 ▲
Niels Bohr

Bohr proposed that the electron orbits the proton like a satellite orbits the Earth, with the necessary force being provided by the attractive nature of the electrical charges of opposite sign on the electron (negative) and proton (positive). From the value of this attractive electrical force he was able to calculate the radius of orbit of the electron, and also its energy, in its most stable, or **ground**, state. Just as satellites can orbit at different distances from the Earth, Bohr reasoned that the electron could be given energy (or **excited**) and exist in less-stable higher energy orbits. By applying quantum theory to his model, Bohr concluded that the electron could only exist in certain **discrete** or **quantised** orbits. This means orbits having distinct, fixed amounts of energy.

As a particular value of energy is associated with each orbit, we say that the electrons have discrete or quantised **energy levels**. When an electron drops from a higher energy level to a lower level, it gives out the energy difference in the form of **one quantum** of radiation, hf. Thus, when an electron drops from energy level E_2 to a level E_1:

$$hf = E_2 - E_1$$

Worked example

Use the data in Figure 16.9 to answer the following questions.

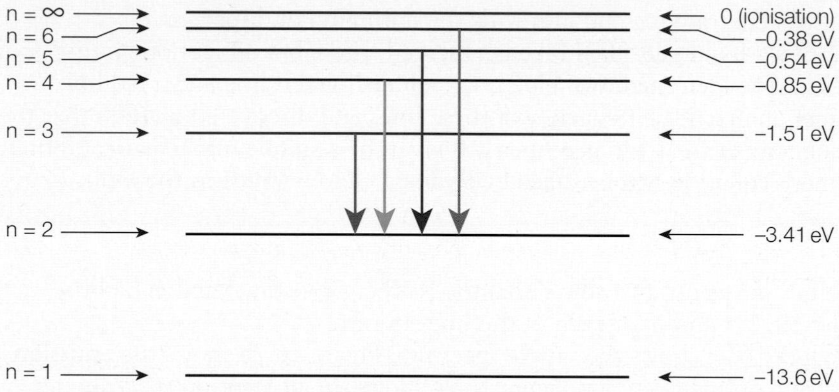

Figure 16.9 ▲

1 Calculate the energy in joules that would have to be supplied for an electron at the lowest energy level to escape from the atom. (This is called the **ionisation energy**.)

2 What is the wavelength of the spectral line emitted when an electron falls from the –1.51 eV level to the –3.41 eV level? Suggest what colour this line would be.

3 Between which energy levels must an electron fall to emit a blue line of wavelength 434 nm?

4 Without doing any detailed calculations, explain why the radiation emitted when an electron falls to the lowest energy level cannot be seen.

Answer

1 Ionisation energy = +13.6 eV = $13.6\,\text{eV} \times 1.6 \times 10^{-19}\,\text{J eV}^{-1} = 2.2 \times 10^{-18}\,\text{J}$

2 $hf = E_2 - E_1 = (-1.51\,\text{eV}) - (-3.41\,\text{eV}) = 1.90\,\text{eV}$

$= 1.90\,\text{eV} \times 1.6 \times 10^{-19}\,\text{J eV}^{-1} = 3.04 \times 10^{-19}\,\text{J}$

$$f = \frac{3.0 \times 10^{-19}\,\text{J}}{6.63 \times 10^{-34}\,\text{J s}} = 4.59 \times 10^{14}\,\text{Hz}$$

$$c = f\lambda \;\Rightarrow\; \lambda = \frac{c}{f} = \frac{3.00 \times 10^{8}\,\text{m s}^{-1}}{4.59 \times 10^{14}\,\text{s}^{-1}} = 6.54 \times 10^{-7}\,\text{m} = 654\,\text{nm}$$

This is visible red light.

3 $f = \dfrac{c}{\lambda} = \dfrac{3.00 \times 10^8\,\text{m s}^{-1}}{434 \times 10^{-9}\,\text{m}} = 6.91 \times 10^{14}\,\text{Hz}$

$E_2 - E_1 = hf = 6.63 \times 0^{-34}\,\text{J s} \times 6.91 \times 10^{14}\,\text{s}^{-1} = 4.58 \times 10^{-19}\,\text{J}$

$\qquad\quad = \dfrac{4.58 \times 10^{-19}\,\text{J}}{1.6 \times 10^{-19}\,\text{J eV}^{-1}} = 2.86\,\text{eV}$

This could arise from an electron falling from the −0.54 eV level to the −3.40 eV level.

4 The smallest possible energy difference when an electron falls to the lowest energy level is:

$\Delta E = (-3.41\,\text{eV}) - (-13.6\,\text{eV}) = 10.19\,\text{eV}$

This is much greater than the 2.86 eV in part **3**, which gives a blue line, so the frequency of the emitted radiation must be much greater than that for blue light. This means it would be in the ultraviolet region of the spectrum and so would not be visible.

The answer to part **4** of the worked example explains why the equation discovered by Balmer contains the term:

$$\frac{1}{2^2}$$

The '2' represents an electron falling to the second energy level. As the lines produced by electrons falling to the lowest energy level were in the ultraviolet range they could not be seen. By substituting '1' into the first term, the wavelengths of spectral lines in the ultraviolet range could be predicted. It was not long before the existence of these lines was established experimentally by Lyman.

This is a good example of how science develops. Scientists had observed spectral lines for some time, and then Balmer derived a mathematical equation to fit these observations. Bohr proposed a model, based on the quantum theory, that gave a physical explanation of Balmer's equation. Bohr's model predicted further lines in the ultraviolet spectrum, which were subsequently found by Lyman.

The development of the quantum theory led to wide-ranging experiments on solid materials ('solid-state physics') in the 1920s and 1930s. After the Second World War, during which the attention of physicists was diverted elsewhere, Bardeen and Brattain discovered the transistor effect in 1947. With Shockley, they were jointly awarded the Nobel Prize for physics in 1956 for 'their researches on semiconductors and their discovery of the transistor effect'. Their work on semiconductors led to the development of silicon chips, without which we would not have today's computers or your iPod!

It is worth mentioning, in the context of the Second World War, that in 1935 Robert Watson-Watt – a Scottish physicist – was asked by the Air Ministry to investigate the possibility of developing a 'death-ray' weapon using radio waves. Watson-Watt did not create a 'death-ray' weapon, but he did find that his radio transmitters could create an echo from an aircraft that was more than 200 miles away. Hence evolved **radar**, an acronym for **ra**dio **d**etection **a**nd **r**anging (appropriately, a palindrome – a word that reads the same backwards as forwards!). Today, worldwide travel would be virtually impossible without radar (What if the *Titanic* had had radar?) and we would not have had mobile phones or microwave ovens (or police speed traps come to that!).

Another application of quantum emission is in the **laser**. This is another acronym, standing for **l**ight **a**mplification by **s**timulated **e**mission of **r**adiation. In a laser, a flash of light is injected into a suitable material, now usually a semiconductor. This excites some of the electrons in the semiconductor atoms into higher unstable energy levels. When the electrons return to their lower energy levels, they emit photons. As these photons pass through the

Figure 16.10 ▲
a) Sir Robert Watson-Watt and
b) William Bradford Shockley

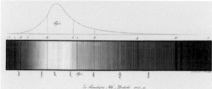

Figure 16.11 ▲
Joseph von Fraunhofer and the
absorption spectrum

semiconductor, they stimulate emission in other atoms, causing an avalanche effect. The light intensity is rapidly amplified and a beam of monochromatic light is emitted; the colour of this depends on the laser material. This beam is **coherent** – the emitted waves are in phase with one another and are so nearly parallel that they can travel long distances. Originally developed for scientific research, lasers are now widely used in industry, medicine and about the home – for example, in CD players and laser printers.

In 1802, Woolaston observed a number of dark lines in the spectrum of the Sun. These lines were accurately mapped by Fraunhofer in 1814 and now bear his name. Fig 16.11 depicts Frauenhofer's original drawing of the **absorption spectrum** of the Sun.

Such a spectrum is called an **absorption** spectrum, because the gases surrounding the Sun selectively absorb certain wavelengths of the Sun's radiation. Electrons in the gas absorb quanta (that is, precise amounts) of energy hf that are exactly sufficient to excite the electrons into higher energy levels. When the electrons drop down to lower energy levels, they re-emit the quanta of radiation in all directions, so the corresponding wavelengths are not observed. This selective absorption of particular wavelengths is an example of **resonance**.

The frequencies of the absorption lines due to a particular element are exactly the same as the frequencies of the emission spectrum of that element. This is shown in Figure 16.7 on page 179. This means we can determine the elements present in the gases surrounding the Sun and other stars and galaxies.

Indeed, the element helium was first discovered in 1868 by observation of a strong yellow emission line in the Sun's spectrum (hence its name, derived from the Greek word for sun, *helios*). A few years earlier the French philosopher Comte had stated: 'There are some things of which the human race must remain in ignorance, for example the chemical composition of the heavenly bodies.' This shows the rashness of prophecy!

16.8 Wave–particle duality

In Unit 1 we saw that light (and, indeed, **all** electromagnetic radiation) has properties associated with waves – for example, polarisation, diffraction and interference. In this section, we have seen that electromagnetic radiation sometimes behaves like particles – for example, the photoelectric effect and atomic spectra. So, is light made of particles or waves? The answer is both!

In Section 10 we saw that electrons, which we think of as particles, could be diffracted and therefore could behave like a wave. Therefore, particles can sometimes behave like waves and waves can sometimes behave like particles! These ideas emerged from the theory of quantum mechanics, which was developed in the early twentieth century. In 1924, de Broglie linked waves and particles by the formula:

$$\lambda = \frac{h}{mv}$$

where λ is the wavelength corresponding to a particle of mass m travelling with a velocity v. Thus was born the concept of wave–particle duality.

Worked example

1 Calculate the wavelength of a football of mass 440 g travelling at 30 m s^{-1} and hence explain why it does not show wave-like properties.

2 a) Show that an electron of energy 15 keV has a wavelength of about 10^{-11} m.

   b) In which region of the electromagnetic spectrum is this wavelength?

Answer

1 $\lambda = \dfrac{h}{mv} = \dfrac{6.63 \times 10^{-34}\,\text{J s}}{0.44\,\text{kg} \times 30\,\text{m s}^{-1}} = 5.0 \times 10^{-35}\,\text{m}$

This wavelength is far too short for us to observe any wave properties of the football.

2 **a)** $15\,\text{keV} = 15 \times 10^3\,\text{eV} \times 1.6 \times 10^{-19}\,\text{J eV}^{-1} = 2.4 \times 10^{-15}\,\text{J}$

$\tfrac{1}{2}\,mv^2 = 2.4 \times 10^{-15}\,\text{J}$

$\Rightarrow v^2 = \dfrac{2 \times 2.4 \times 10^{-15}\,\text{J}}{9.11 \times 10^{-31}\,\text{kg}}$

$\Rightarrow v = 7.26 \times 10^7\,\text{m s}^{-1}$

$\lambda = \dfrac{h}{mv} = \dfrac{6.63 \times 10^{-34}\,\text{J s}}{9.11 \times 10^{-31}\,\text{kg} \times 7.26 \times 10^7\,\text{m s}^{-1}} = 1.00 \times 10^{-11}\,\text{m}$

b) This wavelength is in the X-ray region of the electromagnetic spectrum.

The answer to part **b)** of the worked example explains why electrons can be diffracted by crystals in which the planes of atoms are in the order of 10^{-11} m apart. Wave–particle duality and the de Broglie equation are considered in more detail in Unit 5 of the A-level course.

Three hundred years after Newton and Huygens argued about whether light consisted of particles or waves, we see that both were right. When considering some phenomena, it is more meaningful to think of light as waves and when considering other phenomena it is more meaningful to think of light as particles. This principle of **complementarity** may be one of the most profound contributions of twentieth-century physics.

Let the last word rest with Newton, who, in order to explain the interference effects that he observed, speculated that 'When a Ray of Light falls upon the surface of any pellucid [transparent] Body, and is there refracted or reflected, may not Waves or Vibrations, or tremors be excited in the refracting or reflecting Medium [substance] at the point of Incidence?'

REVIEW QUESTIONS

1 A 60 W reading lamp gives out only 5% of its power as visible light.

a) The visible light intensity at a distance of 1.5 m from the lamp is approximately:

A $0.11\,\text{W m}^{-2}$ B $0.16\,\text{W m}^{-2}$

C $0.42\,\text{W m}^{-2}$ D $2.02\,\text{W m}^{-2}$

b) The energy of a single photon of visible light having a frequency of 660 nm is approximately:

A 1 eV   B 2 eV

C 3 eV   D 4 eV

2 When blue light of wavelength 447 nm is shone onto a sodium surface, photoelectrons are emitted. The work function of sodium is 2.28 eV.

a) The threshold frequency of sodium is:

A $2.9 \times 10^{-34}\,\text{Hz}$ B $1.8 \times 10^{-15}\,\text{Hz}$

C $5.5 \times 10^{14}\,\text{Hz}$ D $3.4 \times 10^{33}\,\text{Hz}$

b) The energy of a photon of the blue light is:

A 1.74 eV   B 2.28 eV

C 2.78 eV   D 2.96 eV

c) The potential difference that would have to be applied to just prevent photoemission would be:

A 0.50 V   B 0.68 V

C 2.3 V   D 5.1 V

3 Hydrogen has energy levels of −0.54 eV and −1.51 eV.

a) When an electron falls from the higher energy level to the lower energy level, the quantum emitted has a wavelength of approximately:

A $1.1\,\mu\text{m}$ B $1.2\,\mu\text{m}$

C $1.3\,\mu\text{m}$ D $1.4\,\mu\text{m}$

b) This wavelength is in which region of the electromagnetic spectrum?

A Infrared   B Ultraviolet

C Visible   D X-ray

4 When a photon of sunlight is incident on a voltaic cell, an electron in the cell gains sufficient energy to move through a potential difference of 0.48 V.

a) What is a photon?

b) Show that the energy to move an electron through a potential difference of 0.48 V is about 8×10^{-20} J.

c) Photons of sunlight typically have energy 4.0×10^{-19} J. Calculate the efficiency of conversion of the energy of the photon.

d) What is the wavelength corresponding to a photon of this energy?

5 A laser pen emitting green light of wavelength 532 nm is shone onto a caesium surface that has a work function of 1.9 eV.

a) What is meant by the term work function?

b) Show that the energy of photons from the laser is about 2.3 eV.

c) What is the maximum kinetic energy of the photoelectrons emitted from the caesium surface? Give your answer in joules.

d) No photoelectrons are emitted when a red laser pen is shone onto the caesium surface. Suggest the reason for this.

6 A lighting manufacturer supplies red, green and yellow LEDs.

a) Which of these will emit photons having the highest energy?

b) The manufacturer's catalogue gives the following data for the red LED:

Wavelength of maximum intensity   630 nm
Visible light emitted   18 mW
Power consumption   120 mW

  i) Show that the photons at the stated wavelength each have energy of about 3×10^{-19} J.

  ii) What is this energy in electron volts?

c) **i)** Assuming that the LED acts as a point source and radiates equally in all directions, show that the intensity at a distance of 30 cm away is about 16 mW m^{-2}.

  ii) Assuming that the diameter of your pupil is 5 mm, approximately how many photons would enter your eye per second at this distance?

  iii) Explain, in terms of photons, why the intensity of the LED gets less as you move further away.

d) What is the efficiency of the LED in converting electrical energy to light energy?

e) Transport departments are beginning to replace filament bulbs in traffic lights with red, green and yellow LEDs. Discuss the advantages of this.

7 **a)** Most physicists believe that light can behave as both a wave and a particle. Name a property of light that shows it can behave as a wave.

b) In 1916, Millikan published the results of an experiment on the photoelectric effect, which proved that light also behaves as particles. He shone monochromatic light onto plates made of different metals. What is meant by the term monochromatic?

c) When light hit the plates, photoelectrons were produced. Millikan found the potential difference that was just large enough to stop these electrons being released. He also investigated how this stopping voltage varied with the frequency of the light used.

Table 16.2 below shows the results of an experiment like Millikan's, in which sodium was used as the metal plate. Plot a graph of V_s against f.

Stopping voltage V_s/V	Frequency of light f/10^{14} Hz
0.26	5.01
0.43	5.49
0.75	6.28
1.00	6.91
1.18	7.41

Table 16.2 ▲

d) The following photoelectric effect equation applies to this experiment:

$$hf = \phi + eV_s$$

where ϕ is the work function of the metal.

What information about the photoelectrons does the value of the term eV_s give?

e) **i)** Use your graph to determine the threshold frequency of sodium.

  ii) Hence calculate the work function, in eV, for sodium.

f) Explain why no electrons are emitted below the threshold frequency.

8 Figure 16.12, which is not to scale, shows some of the energy levels of a tungsten atom.

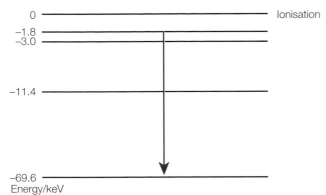

Figure 16.12 ▲

a) An excited electron falls from the −1.8 keV level to the −69.6 keV level. Show that the wavelength of the emitted radiation is approximately 0.02 nm.

b) To which part of the electromagnetic spectrum does this radiation belong?

9 Figure 16.13 shows some of the main components of one type of fluorescent tube.

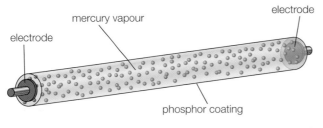

Figure 16.13 ▲

a) When the tube is switched on, a charge flows between the electrodes and the mercury atoms become excited. The mercury atoms then emit radiation.

i) Explain the meaning of the word **excited** as used above.

ii) Explain how the excited mercury atoms emit radiation.

iii) Explain why only certain wavelengths of radiation are emitted.

b) Some of the radiation emitted is in the ultraviolet part of the spectrum. Humans cannot see ultraviolet radiation, so the inside of the tube is coated with phosphor. The atoms of phosphor absorb the ultraviolet radiation and then emit visible light.

i) Suggest why the phosphor emits different wavelengths from the mercury.

ii) A typical classroom fluorescent tube takes a current of 200 mA. Calculate the amount of charge that flows during a lesson that lasts $1\frac{1}{2}$ hours.

iii) Give one advantage and one disadvantage of using a tube such as this rather than a tungsten filament light.

10 The diagrams shown in Figure 16.14 are taken from an explanation of how a laser works. Each diagram illustrates some aspect of a 'two energy level system'. The system consists of an electron in an isolated atom.

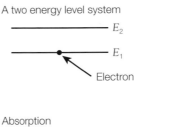

Absorption

Spontaneous emission

Stimulated emission (laser)

Figure 16.14 ▲

a) What is meant by energy level?

b) What is a photon?

c) Write down a formula in terms of E_1 and E_2 for the energy of the photon in the absorption diagram.

d) The laser light emitted by the stimulated emission process must have the same wavelength as the photon in the spontaneous emission diagram. Explain this.

e) The laser light is said to be coherent. Explain the meaning of coherent.

Unit 2 test

Time allowed: 1 hour 20 minutes
Answer **all of** the questions.

For Questions 1–10, select one answer from A–D.

1 When light travels from air into glass, which of the following changes?

A Velocity and wavelength

B Velocity and frequency

C Frequency and wavelength

D Velocity, frequency and wavelength [1]

[Total: 1 mark]

2 When a violin is played, a wave in the string and a sound wave are produced. Which of the following statements is true?

A The sound wave is longitudinal and stationary.

B The sound wave is transverse and progressive.

C The wave in the string is longitudinal and progressive.

D The wave in the string is transverse and stationary. [1]

[Total: 1 mark]

3 A filament reading lamp gives out only 5% of its power as visible light. The light intensity at a distance of 1.25 m from a 40 W lamp is approximately:

A 0.05 W m^{-2}

B 0.10 W m^{-2}

C 0.20 W m^{-2}

D 0.40 W m^{-2} [1]

[Total: 1 mark]

4 The same intensity as in Question 3 could be obtained from a 60 W lamp at a distance of approximately:

A 1.0 m

B 1.5 m

C 1.9 m

D 2.3 m [1]

[Total: 1 mark]

5 Which of the following should **both** have a very low resistance?

A A voltmeter and a laboratory high voltage supply

B A voltmeter and a car battery

C An ammeter and a laboratory high voltage supply

D An ammeter and a car battery [1]

[Total: 1 mark]

6 Power is given by the expression:

A $\dfrac{QI}{t}$

B $\dfrac{QV}{t}$

C $\dfrac{RI}{t}$

D $\dfrac{VI}{t}$ [1]

[Total: 1 mark]

Questions 7 and 8 relate to an electric food blender rated at 240 V, 720 W.

7 The most suitable value of fuse for it would be:

A 1 A

B 3 A

C 5 A

D 13 A [1]

[Total: 1 mark]

8 Its resistance when operating at 240 V is:

A 0.8 Ω

B 8 Ω

C 80 Ω

D 800 Ω [1]

[Total: 1 mark]

9 A student connects two loudspeakers to the same terminals of a signal generator. She walks along a line parallel to the speakers and hears a series of maxima and minima, with a minimum when she is equidistant from the speakers. This is because:

A there is a no path difference between her and the two speakers

B there is a path difference of half a wavelength between her and the two speakers

C one speaker is much louder than the other

D she has connected the speakers in antiphase. [1]

[Total: 1 mark]

10 The central maximum of a single-slit diffraction pattern can be made wider by increasing the:

A frequency

B intensity

C slit width

D wavelength. [1]

[Total: 1 mark]

11 Figure 1 shows a simplified cross-section of part of a DVD.

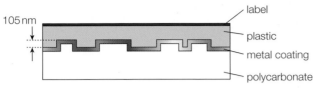

Figure 1 ▲

a) Light of wavelength 640 nm from a laser passes through the plastic coating, which has a refractive index of 1.52. The light is reflected by the flat section and is scattered in all directions if it strikes a bump.

Calculate:

i) the frequency of the light from the laser

ii) the speed of this light in the plastic coating. **[2]**

b) Destructive interference occurs between the light scattered by a bump and that reflected from a flat section.

i) Show that the wavelength of the laser light in the plastic coating is approximately 420 nm.

ii) Hence explain how destructive interference can occur in this situation. **[4]**

[Total: 6 marks]

12 In microwave ovens, the microwaves reflect off the metal walls. The reflected waves interfere and standing waves are set up. This creates hot spots at the antinodes, and cool spots at the nodes, which cause the food to cook unevenly.

a) The manufacturer's label indicates that a microwave oven operates at a frequency of 2450 MHz. Show that the microwaves have a wavelength of approximately 12 cm. **[1]**

b) Figure 2 shows a diagram of the microwave in which two waves reach the point X having travelled different paths.

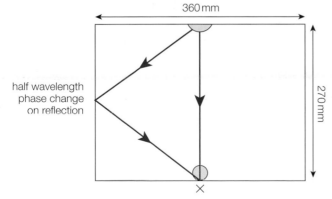

Figure 2 ▲

c) Explain why there is a hot spot at X. **[3]**

[Total: 4 marks]

13 A student wants to light a model house that she has made. She proposes to use an old 9 V mobile phone charger, but her lamps are rated at 3 V. She decides to set up the circuit shown in Figure 3.

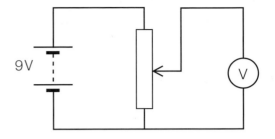

Figure 3 ▲

a) She sets the slider so that the voltmeter reads 3 V. Redraw the circuit diagram and mark the position of the slider with an X. **[1]**

b) She now replaces the voltmeter with one of her 3 V lamps and, much to her dismay, it does not light. Explain the reason for this. **[3]**

[Total: 4 marks]

14 A red light-emitting diode (LED) has the current–potential difference characteristics shown in Table 1.

V/V	I/mA
1.0	0.00
1.1	0.00
1.2	0.00
1.3	0.00
1.4	0.01
1.5	0.05
1.6	0.32
1.7	3.89

Table 1 ▲

a) Discuss whether the LED obeys Ohm's law. **[2]**

b) When the potential difference across it is 1.6 V, calculate:

i) the resistance of the LED

ii) the power dissipated in the LED. **[2]**

c) Suggest an advantage of using a red LED rather than a filament lamp with a piece of red plastic for the rear lamp of a cycle. **[1]**

[Total: 5 marks]

15 Figure 4 shows an arrangement of four 1 kΩ resistors.

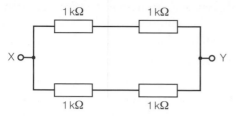

Figure 4 ▲

a) Show that the resistance between points X and Y is 1 kΩ. [2]

b) Suggest an advantage of using such an arrangement rather than a single 1 kΩ resistor. [1]

c) Explain how you could get a resistance of 1.5 kΩ using only 1 kΩ resistors. Draw a diagram of the arrangement. [2]

[Total: 5 marks]

16 The work function for zinc is 4.3 eV.

a) Explain what is meant by work function. [2]

b) Show why it is not possible to observe the photoelectric effect for zinc with visible light of frequency 6.0×10^{14} Hz. [2]

[Total: 4 marks]

17 Polaroid® glasses can be useful when sailing as they reduce the glare from the sea.

a) Explain, with the aid of a diagram, what is meant by polarisation. [2]

b) When sunlight is reflected from a water surface, it is partially plane polarised. Explain how Polaroid® glasses reduce the glare from the sea. [1]

c) At a particular angle of incidence, 53° in the case of water, the reflected light is completely polarised. The refractive index of water is 1.33.

i) Calculate the angle of refraction in the water.

ii) Hence show, with the aid of a diagram, that the reflected and refracted rays are at right angles. Mark all the relevant angles on your diagram. [3]

[Total: 6 marks]

18 In Figure 5, graph X shows how the potential difference across the terminals of a cell depends on the current in the cell. Graph Y is the voltage–current characteristic for a filament lamp.

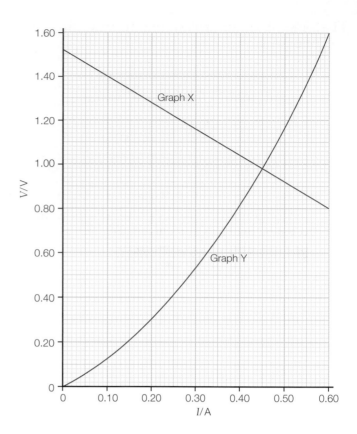

Figure 5 ▲

a) As the current increases, what can be deduced from the graphs about:

i) the internal resistance of the cell

ii) the resistance of the filament lamp? [1]

b) Use graph X to determine:

i) the e.m.f. of the cell

ii) the internal resistance of the cell. [2]

c) When the lamp is connected to the cell, what is:

i) the current in the lamp

ii) the resistance of the lamp

iii) the power developed in the lamp? [3]

d) Draw a circuit diagram of the circuit you would set up to obtain the data for graph Y using two cells for the power supply. [3]

e) Explain

i) why you would need **two** cells

ii) how you would obtain the data. [3]

[Total: 12 marks]

19 a) The current in a copper wire is given by the equation:

$$I = nevA$$

Copy and complete Table 2 by adding the meaning and units for each of the symbols.

Symbol	Meaning	Unit
n		
v		

Table 2 ◄ **[3]**

b) Table 3 gives the values for the resistivity of some common materials at room temperature.

Material	Resistivity/Ω m
Copper	1.7×10^{-8}
Constantan	4.9×10^{-7}
Carbon	1.4×10^{-5}
Silicon	2.4×10^{-3}

Table 3 ◄

Use the equation $I = nevA$ to explain why:

i) silicon has a much greater resistivity than copper

ii) the resistivity of copper increases with temperature while that of silicon decreases with temperature. **[3]**

c) A technician has a reel of 30 gauge (0.315 mm diameter) copper wire.

i) Show that this wire has an area of cross-section of approximately 8×10^{-8} m^2.

ii) What length of wire will the technician need in order to make a 1.0 Ω resistor? **[4]**

d) The technician also has a reel of constantan wire of the same gauge. Explain why it would be better to make the resistor of this wire. **[2]**

[Total: 12 marks]

20 a) Figure 6 shows some of the energy levels for an electron in a hydrogen atom and one possible transition of the electron. This gives rise to a monochromatic spectral line in the red region of the visible spectrum.

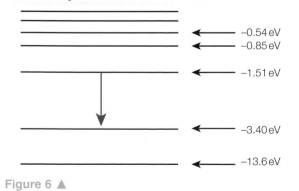

Figure 6 ▲

i) State what is meant by **monochromatic** and explain why this spectral line is monochromatic. **[3]**

ii) Show that the wavelength of this line is approximately 656 nm. **[4]**

b) Quasar 3C 273 was the first quasar discovered and is also the quasar with the greatest apparent brightness. Figure 7 shows the spectrum of light received from 3C 273. The four peaks marked are the shifted spectral lines of hydrogen, which in the laboratory have wavelengths of 410 nm, 434 nm, 486 nm and 656 nm. This is called the Doppler red shift.

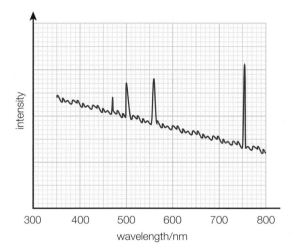

Figure 7 ▲

i) Explain what is meant by the Doppler effect. **[3]**

ii) With reference to the data provided, explain why this is called a red shift. **[1]**

iii) What can be deduced about the motion of 3C 273 relative to Earth from these observations? **[1]**

[Total: 12 marks]

[TOTAL: 80 marks]

Answers to Review Questions

1 Quantities and units

1 a) A
  b) C
  c) D
  d) D

4 a) Diagram should show a distance of 11.2 km with direction S 27° E
  b) Average speed = 3.0 km h^{-1}
     Average velocity = 2.2 km h^{-1} in the direction S 27° E

5 Horizontal component = 120 cos 30 = 104 m s^{-1}
  Vertical component = 120 sin 30 = 60 m s^{-1}

2 A guide to practical work

1 a) For example:

	l/mm	w/mm	t/mm
	95	62	43
	96	62	43
Average	95(.5)	62	43

Table A.1 ▲

Volume: $V = lwt$
= 9.55 cm × 6.2 cm × 4.3 cm
= 255 cm^3

Density = $\dfrac{m}{V} = \dfrac{250\,g}{255\,cm^3}$

= 0.98 g cm^{-3} (980 kg m^{-3})

b) Percentage uncertainty in l
= $\dfrac{1\,mm}{95.5\,mm}$ × 100%
= 1.0%

Percentage uncertainty in w
= $\dfrac{1\,mm}{62\,mm}$ × 100%
= 1.6%

Percentage uncertainty in t
= $\dfrac{1\,mm}{43\,mm}$ × 100%
= 2.3%

The overall percentage uncertainty is therefore in the order of 5% (probably more when also taking into account any manufacturing tolerance and the mass of the wrapper).

This means that the density of the butter could lie between about 0.95 g cm^{-3} and 1.03 g cm^{-3}.
Although the experimental value of 0.98 g cm^{-3} suggests that butter **will** float, the experiment may not be sensitive enough to confirm this beyond all doubt. (You might like to check what happens with a small piece of butter in a cup of water!)

2 a) i) For example, mass of packet of paper,
     M = 2.52 kg
   ii) Mass of single sheet,
     $m = \dfrac{2520\,g}{500} = 5.04\,g$

b) i) For example:

l/mm	297	297	Average: 297
w/mm	210	210	Average: 210

Table A.2 ▲

   ii) Area, A = 0.297 m × 0.210 m
      = 0.0624 m^2
      'gsm' = $\dfrac{5.04\,g}{0.006\,24\,m^2}$
      = 80.8 g m^{-2}
   iii) Percentage difference
      = $\dfrac{(80.8 - 80)\,g\,m^{-2}}{80\,g\,m^{-2}}$ × 100%
      = 1%
      This is acceptable experimental error, particularly when taking into account the mass of the packet and the lack of sensitivity of the kitchen scales.

c) i) For example, measured thickness of packet/mm
     = 48, 47, 48, 49
     ⇒ average = 48 mm
     Thickness of single sheet,
     $t = \dfrac{48\,mm}{500}$
     = 0.096 mm = 0.0096 cm
     Density of paper = $\dfrac{mass}{volume}$
     = $\dfrac{5.04\,g}{29.7\,cm × 21.0\,cm × 0.0096\,cm}$
     = 0.84 g cm^{-3} (840 kg m^{-3})

   ii) The thickness of a single sheet of paper could be checked as follows:
     • First check the micrometer screw gauge or digital callipers for zero error.
     • Fold the paper four times to get 16 thicknesses.
     • Compress to remove any air.
     • Measure 16 t in four different places.
     • Take the average and hence find t.

3 a) For example:

10d/mm	10d/mm	mean d/mm
203	203	20.3

10t/mm	10t/mm	mean t/mm
16.5	16.5	1.65

Table A.3 ▲

Note that d and t have been found by measuring the length of ten coins in a row and the height of ten coins, respectively.

b) $V = \dfrac{\pi d^2 t}{4}$

= $\dfrac{\pi × (2.03\,cm)^2 × 0.165\,cm}{4}$

= 0.534 cm^3

Density = $\dfrac{mass}{volume}$

= $\dfrac{3.56\,g}{0.534\,cm^3}$ = 6.7 g cm^{-3}

c) i) Percentage difference

$$= \frac{(7.8 - 6.7)\,\text{g cm}^{-3}}{7.8\,\text{g cm}^{-3}} \times 100\%$$
$$= 14\%$$

ii) Percentage uncertainty in $10d$

$$= \frac{1\,\text{mm}}{200\,\text{mm}} \times 100\% = 0.5\%$$

Percentage uncertainty in $10t$

$$= \frac{0.5\,\text{mm}}{16.5\,\text{mm}} \times 100\% = 3\%$$

The fact that the percentage difference between the experimental value for the density of the coins and the given value for the density of mild steel differs by 14%, which is much more than the experimental uncertainty, suggests that the coins are **not** made of mild steel. However, the true value for the average thickness is considerably less than that measured. When 10 coins are stacked on top of each other, the thickness measured is actually the thickness of the **rim** of the coin and **not** its average. A better average value could be obtained by using a micrometer and measuring the thickness in several places.

d) The percentage difference between the density of brass $(8.5\,\text{g cm}^{-3})$ and that of mild steel $(7.8\,\text{g cm}^{-3})$ is:

$$\frac{(8.5 - 7.8)\,\text{g cm}^{-3}}{8.15\,\text{g cm}^{-3}} \times 100\% = 9\%$$

This is more than the estimated experimental uncertainties, so it should be possible to distinguish between the two types of coin. In particular, the difference in ten thicknesses would be $(16.5 - 15.2)\,\text{mm} = 1.3\,\text{mm}$, which can easily be detected with a rule. If only **one** coin were available, the difference in thickness would be $0.13\,\text{mm}$, which could be detected easily with a micrometer or digital callipers.

4 a) If $h = \frac{1}{2}gt^2 \Rightarrow t^2 = \frac{2h}{g} = \frac{2}{g} \times h$

A graph of t^2 against h therefore should be a straight line through the origin of gradient equal to $\frac{2}{g}$.

b)

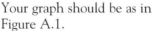

h/cm	40	60	80	100	120
t/s	0.30	0.38	0.42	0.47	0.51
t²/s²	0.090	0.144	0.176	0.221	0.260

Table A.4 ▲

Your graph should be as in Figure A.1.

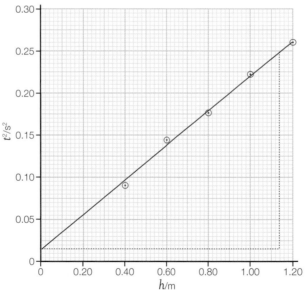

Figure A.1 ▲

c) The graph is a straight line but does not pass through the origin. This suggests a small systematic error. There has possibly been a systematic error in measuring the distance, h (perhaps caused by measuring to the wrong place each time) or, more likely, there has been a systematic error in t due to time delays in releasing the sphere or opening the trap door.

d) Gradient $= \dfrac{0.250\,\text{s}^2 - 0.015\,\text{s}^2}{1.14\,\text{m} - 0.00\,\text{m}}$

$= 0.206\,\text{s}^2\,\text{m}^{-1}$

$\dfrac{2}{g} = 0.206$

$\Rightarrow g = \dfrac{2}{0.206} = 9.7\,\text{m s}^{-2}$

e) Percentage difference

$= \dfrac{(9.8 - 9.7)\,\text{m s}^{-2}}{9.8\,\text{m s}^{-2}} \times 100\%$

$= 1\%$

f) i) The values of h have been recorded only to the nearest centimetre when they would have, presumably, been set to a precision of 1 mm. The values should therefore have been recorded as 40.0 cm, 60.0 cm, etc to reflect this.

ii) Repeat timings should definitely have been recorded, with at least three values for each height. More readings, with larger values of h if possible, would also improve the results.

3 Rectilinear motion

1 a) C   b) B

2 D

3 B

7 a) $v = 12\,\text{m s}^{-1}$

   b) $s = 160\,\text{m}$

9 a) 24 m   b) $22\,\text{m s}^{-1}$

10 a) 7.3 m   b) 1.2 s

    c) $7.6\,\text{m s}^{-1}$ (downward)

11 a) Vertical component = $10\,\mathrm{m\,s^{-1}}$
   Horizontal component = $18\,\mathrm{m\,s^{-1}}$
  b) $t = 2.0\,\mathrm{s}$ c) $x = 36\,\mathrm{m}$

13 a) $-0.2\,\mathrm{m\,s^{-2}}$ b) $1800\,\mathrm{m}$
  c) $10\,\mathrm{m\,s^{-1}}$

4 Forces

1 a) A b) B

2 a) C b) B

5 Work, energy and power

1 C

2 B

3 C

4 A

6 Fluids

1 a) D b) A
  c) D d) C

2 a) $1.5\,\mathrm{kg\,m^{-3}}$

3 $0.76\,\mathrm{m}$ of mercury

4 a) $5.1 \times 10^{-2}\,\mathrm{m^3}$

7 a) $v \approx 1\,\mathrm{mm\,s^{-1}}$

7 Solid materials

1 a) C b) B c) D d) B

8 Nature of waves

1 B

2 A

3 C

4 C

7 a) $25\,\mathrm{Hz}$

  b) Upper trace: $2.0\,\mathrm{cm}$
    Lower trace: $1.0\,\mathrm{cm}$
  c) The oscillations are $\frac{1}{2}$ of a cycle out of phase
    – a phase difference of $\frac{\pi}{2}$ radians.

10 a) i) $0.75\,\mathrm{m\,s^{-1}}$
   ii) $6.0 \times 10^{-8}\,\mathrm{m}$
   iii) $11\,\mathrm{GHz}$

9 Transmission and reflection of waves

1 D

2 A

3 C

4 C

10 Superposition of waves

1 B

2 B

3 D

4 D

9 b) iii) $340\,\mathrm{m\,s^{-1}}$

11 Charge and current

1 a) B b) A

2 a) $16\,\mathrm{kA}$ b) 5×10^{19}

4 a) i) $12\,\mathrm{C}$ ii) 7.5×10^{19}

12 Potential difference, electromotive force and power

1 a) D b) B

2 Power
  Energy, p.d., current
  charge, energy

3 a) $6.5\,\mathrm{A}$ b) $7.8 \times 10^3\,\mathrm{C}$
  c) $1.8\,\mathrm{MJ}$

4 a) $5.5\,\mathrm{J}$ b) $0.72\,\mathrm{W}$
  c) $2.4\,\mathrm{W}$

5 b) $0.95\,\mathrm{W}$ c) $2.4\,\mathrm{mm\,s^{-1}}$
  d) $400\,\mathrm{N}$

6 a) $980\,\mathrm{W}$ b) $125\,\mathrm{A}$
  c) $8\,\mathrm{A}$ d) $2.7\,\mathrm{MJ}$

7 a) £48.40 saving

13 Current-potential difference relationships

1 a) C b) D

2 b) i) $32\,\Omega$ ii) 3%
  c) i) $4.5\,\mathrm{W}$ ii) 10%
  d) i) $32\,\Omega$ ii) $10\,\Omega$

3 b) $177\,\mathrm{mA}$ e) $5\,\mathrm{m}$

4 b) $33\,\Omega$

14 Resistance and resistivity

1 a) B b) C c) D

2 a) Approximately $18\,\Omega$
  b) Approximately $0.60\,\mathrm{V}$

3 a) $0.015\,\Omega$ b) $6.0 \times 10^{-4}\,\Omega$

4 b) i) $5.7\,\mathrm{A}$ ii) $630\,\mathrm{W}$

5 b) i) $0.070\,\Omega\,\mathrm{m^{-1}}$
   ii) $0.3759\,\mathrm{mm}$
   iii) $4.90 \times 10^{-7}\,\Omega\,\mathrm{m}$
   iv) $0.0591\,\mathrm{mm^2}$
   v) $1.08 \times 10^{-6}\,\Omega\,\mathrm{m}$

6 b) 1.1×10^{-6}

7 b) $29\,\Omega$

8 a) i) $50\,\Omega$ ii) $30\,\Omega$
  b) $20\,\mathrm{mW}$ and $300\,\mathrm{mW}$

15 Electric circuits

1 a) A b) D c) D

2 a) $4.7\,\Omega$

3 $18\,\mathrm{A}$

4 b) $0.10\,\mathrm{mA}$ d) $0.10\,\mathrm{mW}$

6 b) ii) $6.4(5)\,\mathrm{V}$ and $3.6\,\mathrm{k\Omega}$
  c) ii) Approximately $2.4\,\mathrm{mW}$

7 a) i) $0.10\,\mathrm{A}$, $0.06\,\mathrm{A}$, $0.04\,\mathrm{A}$
   ii) $0.47\,\mathrm{W}$, $0.0792\,\mathrm{W}$, $0.0528\,\mathrm{W}$

8 b) ii) $106\,\mathrm{k\Omega}$

9 b) i) $22\,\mathrm{k\Omega}$

16 Nature of light

1 a) A b) B

2 a) C b) C c) A

3 a) C b) A

4 c) 19% d) $500\,\mathrm{nm}$

5 c) $7.0 \times 10^{-20}\,\mathrm{J}$

6 b) ii) $2.0\,\mathrm{eV}$
  c) ii) 1.0×10^{12}
  d) 15%

7 e) i) $4.3 \times 10^{14}\,\mathrm{Hz}$ ii) $1.8\,\mathrm{eV}$

9 b) ii) $1080\,\mathrm{C}$

Index